T0382825

BY RICHARD PRESTON

THE DARK BIOLOGY SERIES

The Hot Zone

The Cobra Event

The Demon in the Freezer

Crisis in the Red Zone

ALSO

The Wild Trees

First Light

Micro (with Michael Crichton)

American Steel

The Boat of Dreams

Panic in Level 4

CRISIS IN THE RED ZONE

CRISIS IN THE RED ZONE

The Story of the Deadliest
Ebola Outbreak in History, and
of the Viruses to Come

Richard Preston

RANDOM HOUSE | NEW YORK

Published in the United States by Random House, an imprint and division of
Penguin Random House LLC, New York.

RANDOM HOUSE and the HOUSE colophon are registered trademarks of
Penguin Random House LLC.

Originally published in hardcover in the United States by Random House, an imprint
and division of Penguin Random House LLC, in 2019.

Portions of this book first appeared in *The New Yorker* in somewhat different form.

Grateful acknowledgment is made to the following for permission to use
previously unpublished material:
JEAN-FRANÇOIS RUPPOL: Excerpts from an unpublished journal entitled "Ebola 2" by
Jean-François Ruppol. Reprinted by permission of the author.
NADIA WAUQUIER: Excerpts from an unpublished journal entitled "Ebola Diary" by
Nadia Wauquier. Reprinted by permission of the author.

LIBRARY OF CONGRESS CATALOGING-IN-PUBLICATION DATA
Names: Preston, Richard, author.
Title: Crisis in the red zone: the story of the deadliest Ebola outbreak in history, and of
the viruses to come / by Richard Preston.
Description: New York: Random House, [2019]
Identifiers: LCCN 2019010492 | ISBN 9780812988154 | ISBN 9780812998849 (ebook)
Subjects: | MESH: Hemorrhagic Fever, Ebola—history | Hemorrhagic Fever, Ebola—
epidemiology | Disease Outbreaks—history | International Cooperation—history |
History, 21st Century
Classification: LCC RA644.E26 | NLM WC 534 | DDC 614.5/7—dc23
LC record available at https://lccn.loc.gov/2019010492

Printed in the United States of America on acid-free paper

randomhousebooks.com

4 6 8 9 7 5 3

Book design by Susan Turner

*Dedicated to the brave women and men who risked
or lost their lives to Ebola as they worked at
Kenema Government Hospital protecting
their nation and the world.*

And another angel came out of the temple, crying with a loud voice . . . "Thrust in your sickle and reap, for the time has come for you to reap, for the harvest of the earth is ripe."

—JOHN, Revelation

CONTENTS

PREFACE

*C*risis in the Red Zone is the successor to my 1994 book, *The Hot Zone*. Both books are nonfiction. The people in this story are real and the events are actual, and have been reported and told to the best of my ability. The narrative is supported by hundreds of personal interviews and years of research into published and unpublished documents and source material. Quoted speech comes from my interviews with subjects or from their recollections of words spoken by someone who's no longer alive.

The people in this story have been largely unnoticed by the world. Yet, for me at least, their actions and choices, their lives and deaths, seem to loom at the center of the most destructive rapid epidemic in anyone's lifetime, one which sent feelers into Dallas and New York City, and may be an example of things to come. Though this story focuses on a few people at certain moments in time, I hope it can be thought of as a window that looks at the future of everyone.

Richard Preston

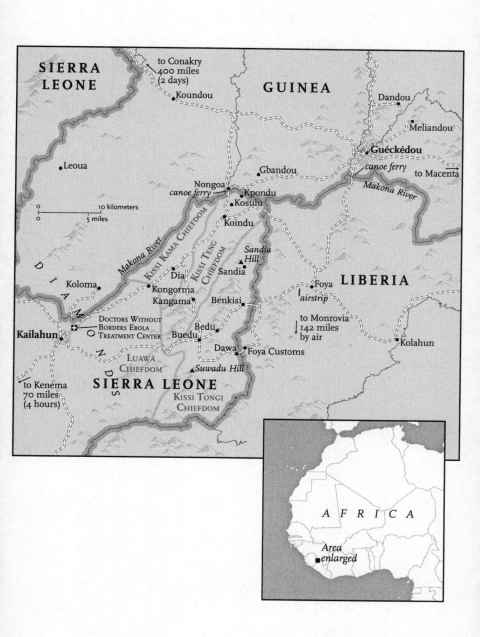

SIERRA LEONE

GUINEA

to Conakry 400 miles (2 days)

Koundou

Dandou

Meliandou

Guéckédou

canoe ferry

to Macenta

Leoua

Gbandou

Makona River

Nongoa
canoe ferry

Kpondu

Kosulu

KISSI KAMA CHIEFDOM

Koindu

Makona River

Sandia Hill

Koloma

KISSI TENG CHIEFDOM

Sandia

LIBERIA

Dia

Foya

Kongorma

Benkisi

airstrip

Kangama

to Monrovia 142 miles by air

Kailahun

DOCTORS WITHOUT BORDERS EBOLA TREATMENT CENTER

Bedu

Buedu

Kolahun

Dawa

Foya Customs

to Kenema 70 miles (4 hours)

LUAWA CHIEFDOM

D I A M O N D S

SIERRA LEONE

Suwadu Hill

KISSI TONGI CHIEFDOM

AFRICA

Area enlarged

SELECTED PERSONS

The Crisis of 2014
(In order of appearance)

ÉMILE OUAMOUNO—a two-year-old boy in Meliandou village, Guinea.

MENINDOR (FINDA NYUMA)—botanist and healer in Kpondu village, Sierra Leone.

DR. HUMARR S. KHAN—chief physician of the Lassa Fever Research Program, Kenema Government Hospital, Sierra Leone.

"AUNTIE" MBALU S. FONNIE—chief nurse of the Lassa ward at Kenema Government Hospital.

LISA HENSLEY—associate director of the Integrated Research Facility (IRF), Fort Detrick, Maryland.

PETER B. JAHRLING—director of the IRF.

SIMBIRIE JALLOH—coordinator of the Lassa research program, Kenema Government Hospital.

DR. PARDIS SABETI—genomic scientist at Harvard University and the Broad Institute.

LINA M. MOSES—scientist with Tulane School of Public Health and Tropical Medicine, New Orleans.

DR. LANCE PLYLER—chief of emergency medical operations for Samaritan's Purse, ELWA Hospital, Monrovia, Liberia.

DR. KENT BRANTLY—chief physician of the Ebola ward, Samaritan's Purse, ELWA Hospital, Monrovia, Liberia.

MICHAEL GBAKIE—biosafety officer and epidemiologist, Lassa research program; deputy to Humarr Khan.

AUGUSTINE GOBA—director of the Lassa Laboratory (the Hot Lab) of the Lassa research program, Kenema Government Hospital.

NADIA WAUQUIER—epidemiologist with biotechnology firm Metabiota.

SAHR NYOKOR—ambulance driver, Kenema Government Hospital.

DR. TOM FLETCHER—WHO doctor; scientist at the Liverpool School of Tropical Medicine, U.K.

LUCY MAY—nurse, Kenema Government Hospital.

IYE PRINCESS GBORIE—nurse, Kenema Government Hospital.

MOHAMED YILLAH—epidemiologist in the Lassa research program, Kenema Government Hospital. Brother of "Auntie" Mbalu Fonnie.

ALEX MOIGBOI—senior nurse, Ebola ward, Kenema Government Hospital.

LARRY ZEITLIN—cofounder and president, Mapp Biopharmaceutical Inc., San Diego.

GENE OLINGER—scientist at the IRF, Fort Detrick, Maryland.

GARY KOBINGER—chief of pathogens at the National Microbiology Laboratory, Winnipeg, Canada.

DR. TIM O'DEMPSEY—WHO doctor; professor at the Liverpool School of Tropical Medicine, Liverpool, U.K.

ALICE KOVOMA—nurse, Ebola ward, Kenema Government Hospital.

NANCY YOKO—nurse and later supervisor of the Ebola ward, Kenema Government Hospital.

DR. JOHN SCHIEFFELIN—WHO doctor; pediatrician with Tulane University School of Medicine, New Orleans.

NANCY WRITEBOL—medical worker with Samaritan's Purse.

The Crisis of 1976
(In order of appearance)

SISTER BEATA (JEANNE VERTOMMEN)—midwife at Yambuku Catholic Mission Hospital, Zaire (now Democratic Republic of the Congo).

FATHER SANGO GERMAIN—priest at Yambuku Catholic Mission, Zaire (now Democratic Republic of the Congo).

DR. JEAN-JACQUES "J. J." MUYEMBÉ-TAMFUN—virologist at the
National University of Zaire, Kinshasa, Zaire.

SISTER MYRIAM (LOUISE ECRAN)—nursing sister at Yambuku
Catholic Mission Hospital.

DR. JEAN-FRANÇOIS RUPPOL—head of the Belgian medical mission
in Zaire.

DR. KARL M. JOHNSON—head of the Special Pathogens Branch of
the Centers for Disease Control, Atlanta, Georgia, U.S.A.

PATRICIA A. WEBB—virologist at the Special Pathogens Branch of
the CDC.

ABBREVIATIONS

Occasionally Used

CDC = Centers for Disease Control

ELWA HOSPITAL = Eternal Love Winning Africa Hospital

HEPA FILTER = high-efficiency particulate air filter

IRF = Integrated Research Facility

NIH = National Institutes of Health

PPE = personal protective equipment

USAMRIID (pronounced "You-SAM-rid") = United States Army
   Medical Research Institute of Infectious Diseases

WHO = World Health Organization

PART ONE

NAMELESS

SACRAMENT

The rains had begun. The nights were clamorous with downpours, and malaria troubled the villages. On the ninth of September, 1976, a woman named Sembo Ndobe arrived at the maternity
ward of the hospital at the Yambuku Catholic Mission, a remote out-
post in Zaire, situated some fifty miles north of the Congo River in a
district of Équateur province called Bumba Zone. Ms. Ndobe had a
high fever, and she was in labor.

The Yambuku Mission Hospital was a collection of one-story pavilions joined by covered walkways, sitting amid African oil palms
and tropical vegetation. The buildings were made of brown bricks, and
had open porticoes running along their sides. The maternity ward was
a modest pavilion with a room that contained nineteen beds. A metal
birthing table, dented and worn, stood at one end of the ward near a
chalkboard where the staff wrote announcements of births. Three
Congolese midwives worked in the ward, along with a Belgian nun
named Sister Beata.

Sister Beata was a middle-aged woman with smooth, dark hair,
which she wore pulled back tightly under a white head covering, and

she had an earthy, warm manner. Her given name was Jeanne Vertommen, and she came from Flanders. In addition to her head covering, Sister Beata typically wore a short-sleeved white blouse and a white skirt. Sometimes, though, either for fun or for practicality, she would wear a long African skirt printed with a bold design. When she worked in the maternity ward, she covered her habit with a cotton surgical gown. She did not wear rubber gloves. Possibly she may have liked the sensation of close contact with babies and their mothers.

Now, she examined Ms. Ndobe. The woman was experiencing agony in her midsection. There was a strange look on Ms. Ndobe's face, a blank, vacant, dazed expression, as if she wasn't all there. She could answer questions, but she didn't seem to be fully aware of her surroundings. The whites of her eyes were inflamed and bright red, and the whites glistened with a film of blood covering the surface of the eyeball. She was bleeding around her teeth, and she may have been urinating blood.

This was nothing very unusual. It looked like a typical case of adult cerebral malaria, or malaria of the brain, a disease which is sometimes called blackwater fever. Blackwater fever causes patients to bleed into their eyeballs, to urinate brown or black blood, and to have hemorrhages from other openings of the body, and it causes brain damage, coma, and death. Sister Beata didn't waste any time trying to diagnose the woman's malady. Her goal was to deliver the woman's child and try to save two lives.

She helped Ms. Ndobe raise and bend her legs, and she inserted her bare hand into the birth canal and checked the dilation of the cervix. She withdrew her hand and saw that her hand and forearm were covered with blood. Ms. Ndobe was hemorrhaging from her womb, and so this seemed to be a troubled birth or a spontaneous miscarriage. Several midwives or nursing aides got Ms. Ndobe onto the metal birthing table in order to help her deliver her fetus or baby. She continued to lose blood from her womb, which ran onto the table.

The aides kept a charcoal fire burning in a hearth outdoors, near the ward, where they heated basins of water. An aide brought a basin of hot water to the birthing table. They dipped a clean towel in the water and placed it around Ms. Ndobe's birth opening, to soften her skin and help ease the seeming agony of her contractions. They also used the towel to mop up the blood that was coming from her birth canal. They rinsed the towel in the basin, to get the blood out of it, got it saturated with fresh hot water, and they gently placed the towel back around the birth opening. They also used the towel to mop blood from the woman's thighs. When the time was right, Sister Beata brought out the child. It was stillborn and covered in blood.

When she saw that the child was dead, Sister Beata might have crossed herself and offered a prayer. The placenta, a mushroom-shaped organ, was a mass of red tissue pressurized with swollen bubbles of hemorrhage. The placenta, however, had not been the only source of Ms. Ndobe's bleeding. After the fetus and placenta were delivered, her bleeding increased. After childbirth, any broken blood vessels in the uterus would quickly seal themselves through clotting, and any bleeding would stop. Ms. Ndobe's bleeding intensified into an uncontrollable hemorrhage pooling on the metal surface of the table. Ms. Ndobe was bleeding out. As her blood spread across the table, her blood pressure fell, her heart began beating rapidly, and her breathing became shallow and irregular. She died of blood loss and shock, either on the birthing table or in one of the beds in the ward. Afterward, Sister Beata probably used some of the hot water to rinse her hands and arms. Hemorrhage during childbirth was a major cause of death in younger women in Africa.

Five days after she delivered the stillborn infant, Sister Beata began feeling strange. A little tired, not quite herself. This feeling continued on for about twelve hours. Then, abruptly, she got a splitting headache and broke with a fever. This was likely malaria. You couldn't avoid malaria in the rain forest regions of the Congo Basin. She went to bed in her room in the community house at the mission. It was a low building that stood not far from the hospital, across from the mission church, a whitewashed structure that seemed to rise like a reef out of a pond of mud that formed around the church in the rainy season. Sister Beata became extremely weak, and she began throwing up. A fierce pain filled her lower abdomen, and she had several episodes of mild diarrhea. The diarrhea was hardly bad, but the pain in her abdomen increased until it became a paralyzing agony, and the pain went into her spine. She became extremely weak, hardly able to move her limbs or get up from her bed.

It was clear that Sister Beata needed to be in the hospital. Some nursing aides carried her out of the community house and placed her in a private room in the women's section of the main adult ward. There, Sister Beata began vomiting into a basin that a nurse held under her mouth as she lay in bed. We do not know exactly what Sister Beata's symptoms were, but judging from the accounts of investigating doctors, who later collected the story of Sister Beata from the surviving nuns at the Yambuku mission, her illness was dramatic and memorable.

She developed projectile vomiting, which is also called rocket vomiting, in which the stomach contracts violently and the vomitus is ejected up to two meters, or six feet, through the air. The vomitus would have ended up on the bed, on the floor, even perhaps on the walls, and certainly on any nurses who were giving her care. On the first day of her vomiting, the vomitus had a normal appearance, but on the second day it came up streaked with blood, or it resembled red paint.

Her rocketing stopped when her stomach was completely empty, and yet her vomiting continued. She started bringing up masses of a black, wet, curdlike material. The substance was recognizable in tropical countries as what is known as the black vomit. The black vomit is a sign of a fatal case of yellow fever. It is a hemorrhage from the lining of the stomach, and it consists of granules or curds of blackened blood which have been partly digested by stomach acid. Black vomit has a characteristic "wet coffee grounds" appearance, in which the granules are mixed with a dark, watery fluid that resembles weak black coffee. She may have come down with hiccups. The hiccups started for no apparent reason and wouldn't stop. Unable to move from her bed, she became incontinent. At first the issuances of stool were whitened with mucus and streaked with blood. As she got sicker, her stool changed into a black liquid. The liquid, known as melena, is hemorrhage coming from the linings of the intestines. The membranes that form the linings of the intestines had died and were now undergoing bacterial decomposition. As the linings of the intestines decayed, blood began leaking from the necrotic tissue, filling the colon with blood. The blood became discolored and, eventually, when the colon was full, the blood was expelled. This was a form of profuse hemorrhage. A rash, consisting of red spots mixed with red bumps, spread around her torso. The red spots, called petechiae, were small hemorrhages occurring inside her skin. Some doctors refer to this type of hemorrhage as bleeding into one's third space. The third space of the body is the soft tissue that lies between the skin and the muscles and fat. The third space can fill up with fluid or blood. Sister Beata's facial expression changed, and her face settled into blank mask, seemingly without emotion, and with inflamed eyes.

YAMBUKU CATHOLIC MISSION
Sunday morning, September 19, 1976

It was still dark when Father Sango Germain, the curate of the Yambuku Catholic Mission, learned that his services were needed at Sister Beata's hospital room. Father Germain was a thin man, in his sixties, with gray hair, glasses, and a long goatee. He gathered up a communion set and a bottle of holy oil, and he left the community house and walked along the path to the hospital. As dawn approached, the jungle surrounding the mission was giving off a wet smell and a piercing whine of cicadas. The palm swifts, small, noisy birds that slept in the palm trees, were beginning to stir, and the outlines of the palms were beginning to appear against the sky, their spidery fronds displaying a slightly menacing appearance.

He walked up the steps of the central pavilion and went to the nun's room, where he found some sisters in attendance, who were praying and keeping vigil. We don't know who among the sisters was there, although there is evidence that one of them was a certain Sister Genoveva. She was a fairly talkative person, less shy than some of the other nuns; she was Flemish, and her given name was Annie Ghyse-brechts.

Father Germain placed his hands on Sister Beata's forehead and prayed. Her fever had passed, and her skin felt merely warm to the touch. A nosebleed had left her nostrils blackened with caked blood. If she had been hiccuping, it had now stopped. She was conscious, but she seemed to be looking at something in the room that nobody else could see. Her eyes were bright red, her eyelids drooped slightly, and delicate pinpoints of blood stood along the rims of her eyelids, like strings of tiny rubies threaded among her eyelashes.

He removed the stopper from the bottle of oil and got some of it on his thumb, and anointed her forehead and hands with the sign of the cross. Soon he moved on to the viaticum, which is a person's last holy communion. He held up the communion wafer.

This is the Lamb of God, who takes away the sins of the world.

She opened her mouth, or possibly he helped her get it open using his fingers. Her lips were dry to the touch. Small black crusts of dried blood had formed in the corners of her lips. Her tongue was bright red and wet, the color of arterial blood, and blood was leaking from her gums. He placed the wafer on her tongue.

Father Germain left no record of what happened next. About a month later, however, Sister Genoveva described the incident to at least two different investigating doctors. According to her, Sister Beata began crying as Father Germain administered the viaticum to her, and tears of blood came from her eyes and ran down her cheeks. The fluid that came from her eyes would have been a mixture of tears and blood. The blood was hemorrhage coming from the wet membranes of her eyelids. Sister Beata's blood had lost its ability to clot, and so it mixed into her tears the way dye spreads into water, and the red liquid ran down her cheeks.

According to Sister Genoveva, when Father Germain saw the bloody tears running down Sister Beata's face, he burst into tears, too. Weeping, he took his handkerchief out of his pocket and gently cleaned the nun's face. Then he dabbed his bloodstained handkerchief into his own eyes, one and then the other, wiping the tears from his eyes, and he put his handkerchief back in his pocket. Father Germain had infected himself with Sister Beata's bloody tears, and he would be dead thirteen days later. Sister Beata died at sunrise on that Sunday morning, just at the moment when the palm swifts were flying out of the palms and beginning a new day.

In the hours that followed Sister Beata's death, the hospital's Congolese nurses began quitting the hospital, and patients began leaving. There was a problem at the hospital. In the main wards, people were dying in beds soaked with blood and feces. There was something demonic about the illness. Their faces went blank, they hiccuped, they had nosebleeds, they became demented, and then, as the disease progressed, they expelled black blood from the anus. Then, for many of these patients, their fever went down and they started feeling better. This was a false dawn, which lasted for about forty-eight hours, and ended with a sudden collapse of blood pressure, known as a crash, followed by death. As the patients crashed and died, they had tremors and shaking at the point of death, and some went into a thrash of seizures at the end. Patients, terrified of the disease in the wards, walked out of the hospital, or if they couldn't walk, their relatives carried them away on motorbikes or on hand-borne Congolese travel chairs. At four o'clock on the afternoon of Sister Beata's death, a nun switched on the mission's shortwave radio and began calling for help.

ROAD'S END

Jean-Jacques "J. J." Muyembé-Tamfun, MD, PhD, vice-dean of the faculty of medicine at the National University of Zaire, sat in the backseat of a Land Rover that was jolting slowly through the jungle along a dirt road in northern Zaire. Night had arrived quickly, and the moon was down. The vehicle's headlights revealed only the road running straight ahead into nothingness between parallel walls of huge trees entangled with vines. Occasional lurches of the vehicle threw J. J. Muyembé toward a doctor from the Congolese army named K. Omombo, who was sitting next to him. Muyembé wasn't sure what the "K" in his colleague's name stood for.

The driver steered the vehicle cautiously through small streams that crossed the road at low spots. As the annual rains grew stronger, roads in the Congo Basin went bad. What looked like a puddle in a road could turn out to be a vat of liquid clay that could engulf a Land Rover to the tops of its windows. For Zaire, this was a good road, Muyembé thought. They had been making an excellent ten miles an hour.

Jean-Jacques Muyembé was a virologist—a scientist who studies viruses—with a PhD from the University of Leuven in Belgium. At age thirty-one, he seemed to be at the start of a brilliant career. Muyembé was an energetic man of medium height, neither stout nor lean, with a round face and prominent cheeks. He had a soft voice and a down-to-earth manner, and he tended to squint when something amused him or when he was thinking hard. Muyembé was the head of the university's biological laboratory.

Muyembé and Omombo were heading for Yambuku Catholic Mission, traveling on orders from Zaire's minister of health, with an assignment to try to identify the mysterious disease at the mission and then to stop it, if they could. The doctors were given a ride to Bumba Zone in the cargo hold of a C-130 Hercules transport aircraft operated by the Zairian Air Force. The plane landed at a dirt airstrip in Bumba Ville, a market town on the north bank of the Congo River situated some eight hundred miles upstream from Kinshasa. In Bumba Ville, the doctors transferred several boxes of medical supplies to the Land Rover and set out immediately for Yambuku. They knew that night would catch them on the road, but they wanted to reach the mission as soon as possible.

Now, as the vehicle bumped and lurched its way through the darkness, J. J. Muyembé turned over in his mind what little he knew about the disease. The chief medical officer of Bumba Zone, Dr. Ngoy Mushola, had visited the Yambuku mission a few days earlier, where he had examined Sister Beata and a number of sick nurses and villagers. Dr. Ngoy had done his best to help the patients, though it didn't take him long to realize that there wasn't much of anything he could do for them. He took careful notes and filed a report. The disease was savage, with a "rapid evolution to death after a mean of three days," he wrote. As he studied the signs and symptoms, Dr. Ngoy Mushola came to believe that he was seeing an unknown disease—a disease that had never been described by doctors and didn't have a name.

J. J. Muyembé wasn't so sure. In medicine, in science, the simplest explanation for something is usually the right one. Just on the basis of probability and common sense, Muyembé thought that the disease was unlikely to be something unknown. Much more likely, it was reasonably common and already had a name. He had no information about whether or not the disease at the mission was contagious.

At eight o'clock, the Land Rover came to a crossroads in a village called Yandongi, and the driver turned left. The road got worse, and entered a tract of deep lowland rain forest, as dark as pitch and inexpressibly wild. The forest wasn't a wilderness. Tens of thousands of people lived in villages scattered in the area. The people were called the Budza, and they hunted and fished in the forest and grew crops of cassava and plantains. The Land Rover slowed to a crawl, and the enveloping forest seemed to become an invisible force pressing in on the vehicle from all sides, like water in an oceanic deep. Muyembé was feeling very thirsty. Eventually the road came out of the forest and began running through cassava fields and groves of oil palms. They had reached the Yambuku mission.

As soon as the Land Rover's headlights washed over the community house, a group of nuns and priests hurried out of it. They were Belgians, and they were carrying a Coleman gas lantern and a pitcher of water. They greeted the doctors warmly and offered them cool water to drink. Muyembé drank deeply and asked for an immediate tour of the hospital.

The sisters were shy women, not ready for much talk, except for the nun named Sister Genoveva, who was less reserved than the others. We can imagine that she held the gas lantern and did much of the talking as the group walked toward the hospital. The group went up a set of steps into the main pavilion.

The building was quiet and dark. Past the entrance, there was an office on the right. They shone the lantern into the office, but there was nothing of interest to be seen. Past the office, the building divided into wings, which had in them open wards filled with rows of iron-frame beds. The beds were empty. The wards were deserted. The floors were filthy in some places, and sheets had been stripped off beds, revealing mattresses made of thin slabs of foam. Many of the foam mattresses appeared to be soaked with body fluids and blood.

This was an abandoned hospital, a shocking sight. J. J. Muyembé had never seen an abandoned hospital in Equatorial Africa. African hospitals were always busy places, crowded with patients and their families and filled with activity—crying babies, people milling around, nurses discussing things, smoke drifting, food vendors walking around offering food for sale. The Yambuku hospital had been treating six thousand to twelve thousand patients a month, but now it lay deserted. So far, Muyembé and Omombo hadn't seen any person with the disease.

The nuns led Muyembé and Omombo along a footpath to a small building that stood in a patch of grass surrounded by undergrowth and palms. The building, made of brown bricks like the others, had frosted windows and a roof made of clay tiles. A concrete ramp led upward to a pair of wooden doors. They opened the doors and found themselves in the hospital's operating theater—a one-room structure.

A modern operating table stood in the center of the room. A surgical lamp had been placed near the operating table. Scattered on the floor were strips of soiled bandage and blood-soaked surgical sponges. The operating room seemed to have been hastily abandoned after a surgery. The group moved on.

As they were going along a covered walkway, they began to hear high-pitched sounds coming out of a doorway. It seemed that the hospital wasn't completely deserted after all. They went through the door into a dark ward, and their flashlights and the gas lantern revealed rows of cribs and small beds. This was the children's ward. The sounds were coming from a crib.

Muyembé bent over the crib and saw an infant boy writhing in agony. He couldn't tell what was wrong with the baby. What had happened to the child's parents? Did the baby have the disease? There didn't seem to be anything Muyembé and Omombo could do to help the child. About five minutes after they found the baby, it stopped breathing and gave up its life.

The group visited the maternity ward, where Sister Beata had worked. At least three women had delivered dead or extremely sick babies in the ward. The infants had come out in pools of blood, and the mothers and infants had all died. The ward appeared to have been abandoned suddenly. A long cotton skirt, printed with a bold black-and-red pattern, had been placed, neatly folded, on a table by the wall. A crumpled surgical gown had been flung across the table by the folded skirt, as if a nurse had taken off her gown suddenly and left the room. The surface of the metal birthing table was smeared with dried blood. There were blood-soaked surgical tampons scattered on the floor. Elsewhere in the birthing room the doctors found basins of brown standing water. The water in the bowls had been clean and hot when childbirths were happening in the ward, and the maternity nurses had used the water to make hot compresses. The blood-tainted water had been sitting in the bowls for days, and had gone putrid in the heat.

LIFE FORM

After touring the hospital, Muyembé and Omombo ate dinner in the community house with the fathers and sisters. The meal that night was fufu, plantains, and rice topped with a green sauce made from pounded cassava leaves. Muyembé really liked the dinner. It was the best country food—simple, vegetarian, delicious, good for the body.

The religious people were in a state of anxiety over the deaths of patients and nurses, and they ate quietly. The Father Superior of the mission, Augustin Sleghers, may have been sweating heavily, his eating utensils possibly shaking in his hands. He seemed to be coming down with malaria. Sister Genoveva, the talkative nun, appeared to be in good health. Alcohol was forbidden at the mission, but Father Léon Claes, a sturdy, convivial priest they called Father Léo, occasionally liked to serve to his fellow missionaries a special cocktail of his own making, which he mixed from vermouth, lemons, and raw banana alcohol. Father Léo seemed to be holding steady, possibly with the help of one of his cocktails. The curate of the mission, Father Sango Germain, a thin, older man with a bushy goatee, was subdued at the dinner table. He may have been grieving or traumatized, because he had given the last rites to Sister Beata only four days earlier. Also present at the dinner table was a nun named Sister Myriam. She was a thin, extremely quiet Belgian woman, of middle years, with a narrow face and a long, delicate nose. Sister Myriam may have been suffering from a creeping sense of malaise, though if she was feeling unwell she didn't mention it at the table.

As he ate cassava leaves and made a bit of conversation with the clerical people, Muyembé thought about what he'd just seen in the empty hospital. He faced two simple questions: What is it? How can it be stopped?

He still hadn't examined any patients with the disease. How could he and Dr. Omombo deal with it if they didn't know what it was? And how could they know what it was if they hadn't seen any people with the disease? As for the dying baby in the children's ward, a baby in rural Africa could die of almost anything. So far, the disease was only a rumor or a mirage, a horripilant shadow that had passed through the hospital, killing patients, and had now gone elsewhere, at least for the moment. Muyembé, as a virologist, had to consider the possibility that this could be a virus. He respected viruses for their power over human lives.

A virus particle is a very small capsule made of proteins locked together in a mathematical pattern. The pattern of the interlocking proteins in a virus is far more complicated than a snowflake. The protein capsule is sometimes wrapped in an oily membrane. Inside the capsule there is a small amount of DNA or RNA, the molecules that contain the genetic code of the virus. The genetic code is the virus's operating system, or wetware, the complete set of instructions for the virus to make copies of itself. Unlike a snowflake or any other kind of crystal, a virus is able to re-create its form. It would be as if a single snowflake started copying itself as it falls, and those copies of the snowflake copy themselves, creating ever-growing numbers of identical copies of the first snowflake, until the air is filled with falling snow, and each flake is a perfect replica of the first flake.

Many virologists feel that viruses are not truly living things. At the same time, viruses are obviously not dead. Virologists like to describe them as life forms. The term is a contradiction: How can something be a form of life that isn't alive? Viruses carry on their existence in a misty borderland that lies between life and death, a gray zone where the things we encounter are neither provably alive nor certainly dead.

One way to understand viruses is to think about them as biological machines. A virus is a wet nanomachine, a tiny, complicated, slightly fuzzy mechanism, which is rubbery, flexible, wobbly, and often a little bit imprecise in its operation—a microscopic nugget of squishy parts. Viruses are subtle, logical, tricky, reactive, devious, opportunistic. They are constantly evolving, their forms steadily changing as time passes. Like all kinds of life, viruses possess a relentless drive to reproduce themselves so that they can persist through time.

When a virus starts copying itself strongly and rapidly in a host, the process is called virus amplification. As a virus amplifies itself in its host, the host, a living organism, can be destroyed. Viruses are the undead of the living world, the zombies of deep time. Nobody knows the origin of viruses—how they came into existence or when they appeared in the history of life on earth. Viruses may be examples or relics of life forms that operated at the dawn of life. Viruses may have come into existence with the first stirrings of life on the planet, roughly four billion years ago. Or they may have arisen *after* life started, during the time when single-celled bacteria had already come into existence— nobody knows.

As for the possibility that a virus might be amplifying itself in people at the mission, J. J. Muyembé thought that he and Dr. Omombo could be seeing an explosive outbreak of yellow fever. Yellow fever, a severe and sometimes fatal disease caused by the yellow fever virus, has a high fatality rate. The fatal form of the disease is known as fulminating visceral yellow fever. A person dying of yellow fever has a high fever, excruciating abdominal pain, and can have black vomit. The virus, as Muyembé well knew, is transmitted from person to person through the bites of mosquitoes. The mosquito, in turn, has caught the yellow fever virus by biting a person who is infected with it. Yellow fever tends to erupt in small communities in the tropics, places like Yambuku, where the virus can amplify itself in the local population like a fire going out of control.

Or, he wondered, could this be a fulminating outbreak of typhoid fever? Typhoid fever is a gastrointestinal infection caused by a type of bacteria—not by a virus. Typhoid fever is extremely contagious, but you can catch it only by eating food or drinking liquid that is contaminated with typhoid bacteria. You can't catch typhoid by contact with blood or body fluids of a typhoid-infected person or by breathing the air near the person. There is extreme abdominal pain, violent, bloody diarrhea, and the bacteria can get into the bloodstream, causing septic shock and death. The victim, in septic shock, can have hemorrhages that flow from any or all of the openings of the body. Typhoid fever has a high fatality rate if it isn't treated with antibiotics.

A good way to diagnose typhoid fever is to collect a blood sample from the sick person and place a few drops of the blood on a petri dish. If there are any typhoid bacteria in the blood, the bacteria will grow in the petri dish, forming a splotch-like colony. Therefore, Muyembé thought, he would definitely want to collect some blood samples from sick people. But he hadn't seen any sick people. No patients to examine, no blood to collect. He and Dr. Omombo were given beds in a guesthouse, where they passed an uneventful night.

The next morning Muyembé and Omombo learned that one of the hospital's Congolese nurses had died in her home during the night. Muyembé immediately made preparations to examine the body. A cadaver can tell you a lot about a disease. He got a blood-collection kit and some glass blood tubes from his boxes of medical supplies, then walked across a soccer field to a group of small houses where the hospital's staff lived. The family of the dead nurse allowed him to enter their house. The house was small, clean, modest. The family had covered the deceased individual with a cloth. At last he would see the disease. Muyembé pulled back the cloth, and felt a sense of shock.

KNIFE

It was a young woman. He hadn't expected to see someone so young, and he perceived her as beautiful, even in death. The sight filled him with a sense of unrecoverable loss. According to a journal left by one of the doctors who later investigated the outbreak, the young woman's name was Amana. She had been working as a nurse's aide, and was possibly new to her job. In the eyes of J. J. Muyembé, she was a colleague, a medical professional, cut down in the course of her work at the hospital. She was a casualty in the field of medicine, a sudden death in a remote place that could not afford to lose her. What might this young woman have accomplished if she had lived a full life? What good might she have done for her patients, what might she have become?

He bent over the body and began his examination. He saw small amounts of sticky, drying blood crusted or smeared around her nostrils and mouth. It was a sign of epistaxis—hemorrhage from the nose. What did her bloody nose mean? What about the blood on the edges of her lips? Had she been having black vomit? Could the black liquid have filled her mouth, smeared her lips, and gotten up into her nose as she vomited?

Yellow fever virus attacks the liver. As the liver fails, the eyes turn yellow or brownish. Indeed her eyes were discolored, either reddish brown or purplish. He wanted to inspect the eyes more closely. In his haste to reach the bodies he had forgotten to bring along rubber gloves. It didn't matter, since the yellow fever virus isn't infectious in direct contact with blood or body fluids—you can only catch yellow fever from the bite of a mosquito.

Using his bare fingertip and thumb, Muyembé lightly pinched an eyelid and rolled it back. The eyelid membranes were red. Inflamed. What did this mean? The appearance of the eyes was not inconsistent with yellow fever.

Gently he drew the legs apart and saw small amounts of blood smeared around the vagina. The blood was sticky and dark. It was hemorrhage, and it resembled the blood that was caked around the nostrils and mouth. The sight pained Jean-Jacques Muyembé beyond words. It was a terrible moment for him.

The woman had been bleeding from her natural openings. Was this typhoid? Was it visceral yellow fever? In order to diagnose yellow fever, he could take a sample of liver tissue and examine it using a microscope. If the woman had died of yellow fever, the tissue of her liver would show distinct changes, and these changes would provide a sharp diagnosis of yellow fever. Unfortunately, though, he hadn't brought a microscope with him to Yambuku. There was no microscope at the mission hospital, at least none he knew of. Therefore he would need to bring a piece of the woman's liver back to his lab in Kinshasa. He would have to cut into the body in order to collect a piece of the liver.

Not only had he neglected to bring rubber gloves with him, he hadn't brought a scalpel, either. But he needed a way to obtain a sample. He was in a hurry, so rather than go back to the hospital and get a scalpel, he fished around in his pocket and brought out his pocket knife. He unfolded the blade; it seemed long enough to reach the liver.

He ran his fingertips across the right side of the cadaver's upper abdomen, just under the rib cage, feeling the shape of the liver, and he located what he thought was probably the middle of the liver. He touched the tip of his knife to the skin at this spot, then pushed the knife straight in.

The blade sank through the skin and abdominal muscles and went into the liver. Immediately blood began pouring out of the incision around his knife. The blood flowed steadily and smoothly, without pulsation. There was no heartbeat; the blood was draining out of the body by gravity. It was good to see a large amount of blood coming out of the incision. It told him that his knife had pierced one of the major veins in the middle of the liver, and the cut vein had let loose a good gush of blood. So he seemed to be on target.

He turned the blade in a small circle, cutting a plug out of the liver. As he worked, blood continued to pour out of the hole in the skin, and it ran down the handle of his knife and began dribbling over his fingers. It had a brownish color and a slippery consistency. There were no clots in the blood. Cadaveric blood doesn't coagulate. The clot-free blood covered his hand and spidered over his wrist and collected along the knob of his wrist bone. Luckily he was wearing a short-sleeved shirt, since the blood would have soaked into the cuff of a long-sleeved shirt. The blood dripped from his wrist, making dots on the floor.

Squinting, as he often did when he was concentrating, Muyembé wiggled his knife until he had gotten a piece of tissue completely detached from the liver. Working the tip of the knife, he began coaxing the piece out of the incision. He took up a red-top blood tube, removed the stopper, and placed the mouth of the tube against the incision. Then he teased the piece of liver out of the hole using his knife. The bit plopped into the tube along with some blood. He plugged the tube with the rubber stopper. Then he pulled the cloth back over the body.

His right hand and wrist were covered with cadaveric blood.

He thanked the family and offered his condolences, went outdoors and found a water pump, and washed the blood off his hand.

And then he got word that there was another nurse with the disease. She was still alive. When he visited her house, he found that she was desperately sick, near death, and she was pregnant. He wanted to get a sample of blood from a living patient, which he could analyze in his Kinshasa lab. He opened his blood kit, placed a rubber band around the woman's arm, found a vein, inserted a blood-draw syringe, and filled a red-top tube with her blood. Afterward, he pressed a cotton ball on her arm in order to stop any bleeding from the needle puncture. But rather than stop the bleeding, the cotton ball became soaked with blood; the pregnant woman's blood wasn't clotting.

He had never seen a patient have an uncontrolled hemorrhage from a needle stick. This was an anomaly.

He returned to the hospital with the tube of blood from the pregnant woman and the tube holding the piece of liver. By then, word had gone around the local villages that doctors had arrived, and people had begun showing up at the hospital seeking help. Dr. Omombo had started triage—caring for as many patients as he could—and Muyembé joined him. The sick patients were being carried to the hospital on litters or on Congolese carry-chairs, or the patients were arriving seated on the backs of motorbikes, clinging to the driver.

Outdoors, in front of the hospital's main pavilion, Muyembé began taking samples of blood from people's arms. He filled about twenty glass tubes with blood. Some of these people, like the pregnant nurse, were hemorrhaging from the stick of a needle, and their blood wouldn't clot. As he drew blood and tried to stanch the leaking from needle sticks, Muyembé got smears of blood on his hands. He put on rubber gloves for some of the messier blood draws, but he was in a hurry. At times he washed his hands with soap, but at other times he didn't wash them at all.

By now it was midday. The liver and the blood samples would start decaying immediately in the tropical heat. He wanted to get these samples back to his lab in Kinshasa for analysis as soon as possible, but Kinshasa was eight hundred miles away, down the Congo River. He needed to keep the samples cold, especially the piece of liver. If the piece of liver was rotten by the time it arrived in Kinshasa, it would be useless for microscopic examination to get a diagnosis of yellow fever. But there was no ice at Yambuku Catholic Mission, no way to chill or freeze a piece of liver.

Muyembé and Omombo made plans to leave the mission by early

afternoon. They would get their things and blood samples into the Land Rover and then head for Bumba Ville. There they would try to get on board a plane that could take them toward the capital. Muyembé packaged the blood tubes and the tube of liver carefully inside a box, to keep the glass from breaking.

Just as the doctors were about to leave, one of the nuns shyly approached J. J. Muyembé. She was Sister Myriam, the thin nun with a narrow face and a long, bony nose. Her given name was Louise Ecran. She had been working in the hospital as a nurse. "I have a fever and a headache," she said quietly to Muyembé.

He asked the nun if it would be all right to give her a brief physical exam. She agreed, and they went into a private room.

FLIGHT

Sister Myriam had undressed herself above the waist. Her arms were slender, and on her wrist she wore a small, elegant wristwatch. As he examined her, Muyembé observed a strange rash covering her breasts and torso.

The rash consisted of a carpet of red goosebumps rising out of a splotchy, speckly reddening of the skin. The red speckles or splotches were small bleeds, called petechiae, appearing beneath the skin's surface. They were tiny star-shaped pools of blood, which were spreading from leaky blood vessels into the underlayers of the skin. The bleeds were easily visible through the nun's translucent European skin.

Muyembé had never seen a rash like this. It was another anomaly. It now seemed to him that he was seeing a polymorphic disease. This is a disease that takes different forms in different people at different stages of the illness. A polymorphic disease is difficult to recognize because it is a shape shifter, a disease with many faces. To try to see the whole shape of the Yambuku disease was like looking at reflections of sunlight moving on restless water, and seeing only a dance of ever-changing flashes that never coalesce into an image of the sun.

Muyembé spoke to the nun gently, in French. "I think, Sister—
I think that we must go to Kinshasa to get answers, because I don't
understand the nature of this disease."

She refused to leave the mission. "I can't go to Kinshasa with you,"
she said. "If I go to Kinshasa, I will abandon my post and my work."

She had become his patient. A patient has freedom of choice. All he
could do was reason with her. "It is important that we go to Kinshasa,"
he said, "because there we have laboratories, and there we can find
solutions to the mystery."

"I cannot go. The people would say I had abandoned them."

He offered to take her to Ngaliema Hospital in Kinshasa. It was a
private hospital situated in the old colonial district—the best hospital
in the city. If he and the doctors at the hospital could identify the disease, then they would be able to offer treatment for it. And this would
benefit all the people of Yambuku. If she let him take her to the hospital, he said, she would be continuing her medical service to her patients
in Yambuku.

At this point, Sister Myriam agreed to go.

She would need to be accompanied by a female companion, in case
she required intimate care during the journey. One of the nuns, Sister
Edmonda, agreed to go along, and would care for Sister Myriam.

Then, just as they were about to leave, the superior of the mission,
Father Augustin Sleghers, told Muyembé and Omombo that he, too,
had a fever. The doctors welcomed the priest to come along. A commercial aircraft normally made a scheduled landing on the dirt airstrip at Bumba Ville three times a week. The next flight was due to
land the following evening, in about thirty hours. They would try to
get on that plane, assuming it arrived. The rains were making air
travel uncertain.

Muyembé placed the box of sample tubes in the back of the Land
Rover, the group crowded themselves into the vehicle, and they set out
for Bumba Ville. The town was fifty miles away, a five-hour drive
along the ornery dirt road that had brought the doctors to Yambuku.
The Land Rover was now jammed with six people, two of whom were
showing symptoms of the disease, Sister Myriam and Father Sleghers.

In addition, there was Sister Edmonda, along with Omombo, Muyembé, and the vehicle's driver.

Muyembé ended up seated next to Sister Myriam, pressed against her in the backseat and bumping against her as the vehicle lurched. Her fever seemed to be getting worse. He could feel heat coming off the nun's body, and her face and arms dripped with sweat. He noticed that her strange rash was spreading, too. Now it was emerging from under the collar of her blouse and moving up her neck toward her face. Muyembé also saw that the rash was coming out from under the short sleeves of her white blouse and was spreading downward on her bare arms. One of her arms rubbed against his bare arm, and he could feel her sweat on his skin. Sister Myriam remained stoic as they jolted down the road.

They arrived at the Catholic mission in Bumba Ville after dark. The superior of the mission, Father Carlos Rommel, welcomed the group and got them installed in rooms for the night, and the two nuns went into seclusion. The group spent the next day resting quietly at the Bumba mission. There was no ice at the mission, no way to keep the tubes of blood and the piece of liver cold. The samples were starting to decay.

The following evening, near sundown, a twin-engine turboprop Fokker Friendship, the workhorse of African skies, touched down on the Bumba airstrip. The little group climbed on board, and the doctors helped the nuns and the priest get seated. The Friendship took off and climbed over the river, and then banked toward the east, turning away from Kinshasa, away from the group's final destination, and began following the Congo upstream into the east, along a bearing that would take the plane toward Lake Victoria and East Africa. This was the only flight out of Bumba, so the doctors had had no choice except to take it. The Friendship continued to follow the river upstream, its course bending gradually toward the southeast, while the sun fell below the horizon and the sky deepened to cobalt blue. The Congo stretched ahead of the plane, miles wide and braiding among islands, its multiple channels becoming indistinct in the rising darkness. Father Seghers and Sister Myriam were getting slowly sicker. The infectious agent was traveling on the Friendship, too. Along with the humans, it was headed ultimately for Kinshasa, a city with a population of two million, and with airline connections to cities all over the world.

We now know that the agent was a previously unknown virus that would soon be given the name Ebola. The virus is a member of the filovirus family, and it is a parasite that exists, normally, in some creature that inhabits the ecosystems of Equatorial Africa. This creature is the natural host of Ebola. It may well be a type of bat, or some small animal that lives on the body of a bat—possibly a bloodsucking insect, a tick, or a mite. Occasionally a few particles of Ebola escape from Ebola's natural host and enter the bloodstream of a person. The virus begins replicating in the person's cells.

Ebola multiplies to extreme concentrations in the bloodstream. When a person dies of Ebola, a drop of their blood the size of the "o" in this text can easily contain a hundred million particles of Ebola. Ebola can destroy a person's immune system in seven to ten days. HIV requires years to wipe out a person's immune system. Ebola patients typically become disoriented or deranged: The virus affects the brain in some unknown way, and it causes a change in the person's facial expression, giving the face a masklike appearance. Ebola patients die suddenly, in a cascade of shock, and often, at the point of death, their bodies shake with tremors and seizures. Nobody knows exactly what Ebola does to the human body as it destroys it: When a person dies of Ebola virus disease, the cause of death is unknown.

Despite its ferocity in humans, Ebola is a life form of mysterious simplicity. A particle of Ebola is made of only six structural proteins, knitted together to become an object that resembles a short strand of cooked spaghetti. An Ebola particle is only around eighty nanometers wide and a thousand nanometers long. If an Ebola particle were the size of a real piece of spaghetti, then a human hair would be about twelve feet in diameter and would resemble the trunk of a giant redwood tree.

Experiments suggest that if one viable particle of Ebola enters a person's bloodstream, it can cause a fatal infection. Ebola is transmitted among people through direct contact with liquids that come from the body, especially blood and sweat. Once an Ebola particle enters a person's bloodstream, it drifts until it sticks to a cell. The particle is pulled inside the cell, where it takes control of the cell's machinery and causes the cell to start making copies of it. Most viruses use the cells of specific tissues to copy themselves. For example, many cold viruses replicate in the sinuses and the throat. Ebola replicates in all tissues of the body except for the skeleton and the large muscles of the skeleton, and it has a special affinity for the linings of blood vessels. After about eighteen hours, the infected cell is releasing thousands of new Ebola particles, which sprout from the cell in threads, until the cell has the appearance of a ball of tangled yarn.

Each Ebola particle is studded with about three hundred soft knobs. The knobs help the particle get inside a human cell. Inside each Ebola particle is a tube made of coiled proteins, which runs the length of the particle, like an inner sleeve. Viewed with an electron microscope, the sleeve has a knurled look. Like the rest of the particle, the sleeve has been shaped by the forces of natural selection working over long stretches of time. Ebola's family of viruses, the filoviruses, appear to have been around in some form for millions of years. Within the inner sleeve of an Ebola particle, invisible even to a powerful microscope, is a strand of RNA, the molecule that contains the virus's genetic code, or genome. The code is contained in nucleotide bases, or letters, of the RNA. These letters, ordered in their proper sequence, make up the complete set of instructions that enables the virus to make copies of itself.

By one recent count, an Ebola particle has 18,959 letters of code in its genome. This is a small genome, by the measure of living things. The human genome, for example, has around 3.2 billion letters of DNA code, and the loblolly pine has 22 billion letters of code. Viruses such as Ebola, which use RNA for their genetic code, are prone to making errors in the code as they multiply. These errors are called mutations.

Ebola is one of a class of pathogens known as emerging viruses.

An emerging virus, typically, is one that naturally infects some species of wild animal but is also capable of infecting humans. The virus can jump from its wild host into a person and can begin replicating in the person. This is a process known as the cross-species jump of a virus. According to genomic scientists who study the code of viruses, viruses have been doing cross-species jumps, moving from one kind of host to another, for billions of years. Typically a virus mutates rapidly as it moves into a new kind of host. The virus's genetic code starts changing as it encounters new conditions in a new host. The virus is adapting to its new host, ensuring its survival through the ages.

After making a cross-species jump from an animal into a person, an emerging virus can start moving from person to person, starting chains of infection, expanding its range in its new human host. A virus that is making a cross-species jump out of an ecosystem into people can be thought of as a wild creature in its own right. Like many wild creatures, an emerging virus can be unpredictable and dangerous.

At Yambuku, in 1976, a few particles of Ebola slipped out of an animal that lives in the central African rain forest and got into the bloodstream of one person. The first human victim of Ebola at Yambuku has never been identified. The person may have been a forty-two-year-old schoolteacher at the Yambuku mission named Antoine Lokela, who died in the mission hospital on September 8, 1976, with severe hemorrhages coming from the openings of his body. He gave the virus to his wife, Sophie Lisoke, who broke with Ebola and nearly died, but survived her illness. Sophie Lisoke was the first known survivor of Ebola disease.

From the body of its first victim the virus started moving—ancient, opportunistic, adaptable, cunning in a biological sense. Ebola's only mission was to never stop replicating, and to never stop moving from person to person, and thereby to make itself immortal in the human species.

Nobody knew then, nor does anybody know now, where emerging viruses are going or what one of them might become. The human host has been gathering itself into gigantic supercities, teeming urban megahives packed with tens of millions of individuals jammed into a small space, who are breathing one another's air and touching one another's bodies. The supercities are growing larger all the time. Many of the world's largest supercities are crowded with people who have little or no access to doctors and medical care. The cities are connected by airline routes, and the human host has zero immunity to any emerging virus.

Ebola is roughly as contagious as seasonal flu.

PART TWO

HEAT LIGHTNING

THE WASHING POOL

Thirty-seven years later

UPPER MAKONA RIVER, WEST AFRICA
December 2013

The Kissi people of West Africa, who speak their own language and have their own traditions, live in a green countryside, scattered with hills, which spreads across parts of Sierra Leone, Guinea, and Liberia—three small nations that are grouped together along the West Coast of Africa. The Kissi territory covers an area where the national borders of the three countries converge in a triskelion, or triple spiral. Long stretches of the borders follow the course of the Makona River. A narrow, olive drab river broken by occasional rapids, the Makona winds through Kissi lands and then runs southwest through Sierra Leone until it ends in the Atlantic Ocean. In this book the Kissi lands surrounding the river will be called the Makona Triangle. The people of the Makona Triangle cross the river and travel among the three countries constantly, visiting relatives, doing business, seeking medical care, and they pay little attention to which of the three countries they happen to be in at a given moment.

The Makona Triangle lies at the northern end of a belt of tropical forest and natural grassland that once extended for a thousand miles along the curving coast of West Africa, from Guinea down to Ghana.

The West African forest is an ecosystem of immense biodiversity. It is populated with hundreds of different species of trees, together with many kinds of grasses, shrubs, vines, ferns, and mosses, and it is a home to chimpanzees, fungi, elephants, lichens, antelope, algae, protozoa, monkeys, slime molds, mites, bats, segmented worms, smooth worms, rodents, frogs, birds, insects, spiders, and truly astronomical numbers of bacteria. The West African forest also contains an ocean of viruses.

This ocean of viruses in the world of living nature is known as the virosphere. The virosphere includes all viruses as well as infective proteins called prions. The biosphere, as distinct from the virosphere, is the universe of organisms that are made of cells. The biosphere includes everything alive, from tigers to black slime on rocks. All living organisms in the biosphere are made of cells—single-celled organisms and multicellular organisms alike.

The virosphere and the biosphere exist together and interpenetrate each other, like milk in tea, like mist in air. Everything that lives gets infected with viruses. As far as anyone knows, viruses replicate in the cells of all species of living things, *all* of them, from bacteria to blue whales. The virosphere permeates the earth's atmosphere, which is filled with viruses blowing in the wind. Around ten million virus particles land on every square meter of the earth each day, drifting down from the air. Viruses saturate the soil and the sea. A liter of seawater contains more virus particles than any other form of life. Viruses exist in vast numbers in the human gut, infecting all of the four thousand different kinds of bacteria that live naturally in a person's intestines. Viruses can sometimes infect other viruses. A giant virus named the Mamavirus, which was discovered infecting amoebae that live in a water-cooling tower in Paris, gets infected by a small virus called the Sputnik. A Mamavirus particle with Sputnik disease is one sick virus—deformed and unable to replicate very well.

Almost all viruses in any ecosystem are unknown to science.

In recent decades, much of the West African forest has been cut down. At the same time, the human population has grown dramatically. What were once villages have become small cities, and small cities have become metropolises of millions. As this happened, the forest was steadily eaten away and turned into small fragments. The wild grasslands were cultivated, and the land became a quilt-work of cassava fields, rice fields, plantations of oil palms, groves of cocoa trees, and large tracts of a type of dense, brushy thicket called farmbush. Even so, many small pieces of the ancient West African forest remain, especially on the summits of hills, where a stand of old trees can form a kind of topknot, as if the trees have gathered themselves into a defensive huddle against a besieging enemy.

The Kissi people regard the remaining bits of wild forest as sacred places. The chiefs of villages protect these forest patches, and won't allow anyone to cut down a tree in a sacred patch. Kissi people hold ceremonies and bury their dead there, and the spirits of family ancestors dwell in the forest patches. In biological terms, the fragments of old forest in West Africa are the remains of a primeval ecosystem that has existed for millions of years and is now under threat and disappearing. The remaining fragments of wild forest have become zones of contact where the life forms that still exist in the forest mix with the human world.

In older times, when large tracts of unbroken forest still existed, Kissi hunters pursued antelope, monkeys, duiker, elephants, and buffalo. With the breakup of the forest, the game animals have largely disappeared or become very scarce. These days, Kissi hunters devote themselves to trapping cane rats and shooting bats. The cane rat is a meaty rodent that can grow as big as a raccoon, and it inhabits grassland and farmbush. The most highly prized bats are fruit bats, also called flying foxes. They are large bats with cinnamon fur, large, alert eyes, and a pointed nose, like a fox. A well-aimed shotgun blast into the top of a palm tree can bring down ten or twenty flying foxes at once. The meat of a flying fox is said to have a mild taste. Local people cook it into a sauce that's poured over rice.

There's another kind of bat that Kissi people call the *lolibelo* or flying mouse. Flying mice are small and gray, and have a thin, hairless tail, like a mouse, and they can crawl surprisingly fast. They eat insects rather than fruit. Flying mice stink; they give off an acrid reek that smells like mouse piss. Many Kissi adults refuse to eat flying mice. Kissi children, on the other hand, do eat them. They don't seem to mind the smell as much as adults do.

In a Kissi village called Meliandou, which is in the Makona Triangle in Guinea, the children liked to play at the foot of a dead tree that smelled bad. Meliandou is situated below a deforested hill. It is about five miles from the Makona River and the border with Liberia, and is about fourteen miles from Sierra Leone. The village is a tight group of thirty-one houses, plus a schoolhouse and a tiny medical clinic. The houses are made of mud bricks or concrete blocks, and have metal roofs, stained with rust. Like many Kissi villages, Meliandou is encircled by a small forest, a deep ring of trees. Most of the trees in Meliandou's ring forest have been planted by people, and their harvest is used for food and sold for cash. There are cacao trees, oil palms, and mangos in Meliandou's ring forest—but mixed in among them are a few wild trees, with heavy trunks and magisterial crowns. A stream flows through Meliandou's ring forest and goes down to a pool, where the women of the village wash clothes, bathe, and wash their children. The smelly tree stood near the washing pool. It was tall and hollow, a silver relic of a vanished ecosystem. Kids liked to play around the tree while their mothers were washing at the pool. They would hide behind the tree's fluted buttresses, which came out of the tree in thin blades arranged in a star around the base of the tree, and they liked to crawl inside the tree through a hole in its base. The hole opened up into a cave that ran upward through the center of the tree and out of sight. The tree cave was full of stinky flying mice.

In mid-December 2013, a woman of the village named Sia Dembadouno took her two-year-old son, Émile Ouamouno, and probably her daughter, four-year-old Philomène, down to the washing pool. While the mother was at the pool, it seems that little Émile wandered off with a group of older children and played around the dead tree.

The children of Meliandou would sometimes build a small fire inside the cave at the base of the tree. The smoke would go up through the hollow tree, and the bats would get annoyed and start flying out. Some would get smoke-addled, fall down, and land in or near the fire.

The older children would gather around the hole at the base of the tree, holding sharpened sticks, and they'd spear the bats. They roasted the bats over the fire on their sticks as if they were marshmallows. Unlike some adults, the kids had no problem eating flying mice. They would eat the roasted bats straight off the stick, and they often shared bat-on-a-stick with one another. Émile was a toddler, too small to be able to kill or cook any bats, but he might have tasted a raw or undercooked bat, or he might have played with a groggy bat, or he might have gotten some bat blood or bat urine in his eyes or in a cut in his skin.

Or the little boy might have gotten a bite from a bat fly. A bat fly is a blind, wingless fly that drinks the blood of bats. It has long, jointed, hairy legs, like a spider, and it is a good crawler. Bat flies are found in bat roosts, where bats hang crowded together, and the flies crawl from bat to bat, sucking their blood. Possibly a bat fly crawled onto Émile and bit him. The fly might have had some bat blood in its mouthparts, and might have injected a small amount of bat blood into the boy. The bat blood may have been contaminated with a few particles of a virus. This is all speculation. Nobody actually knows how Émile got infected. All we know is that a few particles of a virus, maybe only one particle of a virus, emerged from the virosphere and entered the little boy.

On Christmas Eve, the boy came down with diarrhea. It turned into a black liquid, and he died on December 28, in his mother's arms. A week after Émile died, his four-year-old sister, Philomène, also got black diarrhea and died. In villages in Africa where houses don't have running water, women often use their bare hands and saliva to clean children who throw up or soil themselves. After Philomène's death, the children's mother came down with a fever. She died on January 11, 2014. She was twenty-five at the time of her death. Her family buried her next to her house, as is the tradition in West Africa. Shortly afterward, her mother, Philomène's and Émile's grandmother, started vomiting and died a few days later.

The village midwife had nursed the children's mother and grandmother while they were sick. Not long afterward, the midwife broke with a fever. By this time the village was starting to get really frightened by the chain of deaths. The midwife's relatives got very worried about her and took her to a hospital in a small city called Guéckédou. The city, with a population of 200,000, is in Guinea. It is seven miles from Meliandou. The midwife died in the Guéckédou hospital. Then a medical worker at the hospital, who'd taken care of the midwife as she died, came down with the sickness. This medical worker decided to seek care at a hospital in a town called Macenta, which is forty miles from Guéckédou. The medical worker died at the Macenta hospital. The illness now began moving around Macenta, as well as moving around Guéckédou. A viral lightning bolt had come out of the forest and struck a little boy. The child had been killed, and he had started a chain of infections in a few more people. The virus started amplifying itself in two places in Guinea, and then it jumped to more places, and soon a viral fire was smouldering in the Makona Triangle.

Months later, after the fire had gotten bigger and had been noticed, an expert in viruses named Fabian Leendertz, who works at the Robert Koch Institute in Berlin, spent eight days in Meliandou village with a team of colleagues, including an anthropologist, in an effort to trace the origin of the virus back into the ecosystem. It came down to a question of exactly how little Émile might have gotten infected in the first place. The boy was the first identified case of the disease—the index case. Apparently the virus had leaked out of some wild animal and gotten into the boy. His body had been the bridge over which the virus had traveled in its passage from the virosphere into the human species. But exactly what kind of creature had the virus come from? Where in the ecosystem did the virus hide?

THE WOMAN WITH
ALMOND EYES

MELIANDOU
February–early March, 2014

The residents of Meliandou, Fabian Leendertz recalls, "really liked it when we came." By the time they showed up, however, nobody in the village was sick. The virus had moved on. But the deaths had upset the villagers deeply, and they really wanted to find out where the disease had come from. The villagers helped Leendertz's team trap bats and rodents so they could see if the virus had come from any of these animals. The villagers also shared a great deal of information about the way they hunted wild animals, about who had died in the village and when, and about their funeral practices.

Just before Leendertz and his team arrived in Meliandou, however, the bat tree somehow caught fire and burned. A large number of dead bats fell out of the tree during the fire, raining down through the hollow tree and landing on the ground around it. The villagers filled six rice sacks full of dead flying mice, cooked the bats more thoroughly, and ate them, despite the smell. Leendertz learned that nobody in the village had gotten sick from direct exposure to or eating the bats. This would suggest that the virus didn't come from the bats. Or it would suggest that only a *few* of the bats carried the virus, while most of the bats did not. That is, the virus could be a rare disease of the bats, a disease that most of the bats never catch. Bats have their own rare diseases, of course, just as humans do. In any event, none of the rodents or other kinds of bats that the team members and villagers collected was carrying a deadly virus. In the end, Fabian Leendertz was never able to prove that bats or any other kind of creature had been the source of the disease. Nevertheless, he had a strong suspicion that the little boy had caught his illness from a bat. "We only have pieces of evidence, but there's no proof," Leendertz says.

In the months after its emergence in Meliandou, the disease kept spreading. In a village called Dandou, which is a fifteen-minute drive by motorbike from Meliandou, a man got sick. He was a relative of the midwife of Meliandou, the woman who'd died in the Guéckédou hospital after she'd taken care of Émile's mother and grandmother. The man realized he was dying, and he asked his family to carry him into a patch of sacred forest. They placed him on the ground under the trees, where he died surrounded by his loved ones. Afterward, following local tradition, the man's friends and family members took turns lying next to his body in the forest. They embraced the body, wept over it, and ate meals next to it during which they served foods that the man had liked. This was a way of remembering the man and cherishing their love for him. The disease attacked some of the mourners afterward.

In late February, in a village called Kpondu, in Sierra Leone, a woman in her thirties named Sia Wanda Koniono decided to travel to a city in Guinea called Kissidougou to visit her son. Kpondu village is a tiny group of houses situated only three hundred yards from the Makona River and the international border with Guinea. Ms. Koniono got across the river in a wooden pirogue poled by ferrymen. On the other side, in a bustling town in Guinea, she either hired a ride-share taxi or got on a jitney bus, which in that area is called a *buda-buda*. During the ride, she sat next to a passenger who was sick. Back in Kpondu village, after visiting her son in Guinea, Ms. Koniono came down with diarrhea and vomiting.

Her sickness got worse, and she sought medical help from a neighbor named Finda Nyuma. Finda Nyuma was a highly respected traditional healer. She was better known by her professional name, Menindor. Working as Menindor, Ms. Nyuma performed exorcisms and gave people medicinal compounds made from plants.

Menindor was a tall, impressive woman, a botanist with a deep knowledge of plants and the spirit world, and she was mysterious in her ways. She had almond-shaped eyes and a long face, and a gentle, enigmatic smile, and she covered her head with a lacework shawl. No one seemed to know Menindor's exact age, but she was neither young nor particularly old.

We wouldn't know about the existence of Menindor or any details about her if not for remarkable epidemiology and investigative reporting done by Dr. Sheri Fink and colleagues of hers at *The New York Times*, who traced the early spread of the virus in Sierra Leone. Menindor was a beloved figure in the villages along the Makona River. Many of her patients were women and young girls, and they came from villages far away to be healed by Menindor. The local people believed that Menindor kept a powerful snake in a chest in her house. This was no ordinary snake, but a supernatural being.

Ms. Koniono, sick after her trip in Guinea, got treatment from Menindor, but her vomiting and diarrhea continued. Finally her relatives decided to take her to the city of Guéckédou, in Guinea, where her brother lived, for medical treatment at the Guéckédou hospital. The city is ten miles from Kpondu village. She started throwing up blood in the hospital, and she died there on March 3. Her family intended to have her buried in Sierra Leone. It is traditional for women to prepare a body for burial. Five of Ms. Koniono's sisters washed her body. Following the usual practice, they gave the body a kind of enema, which purged the contents of the bowels and cleansed the body internally. This is something important to do in a tropical climate, or the body deteriorates rapidly. Then they poured clean water over the body. Ms. Koniono's body was brought back to Kpondu village. A day later, her relatives took her body to a spot a few miles away, and gave her a burial.

In the following weeks, in Guinea, all five of Ms. Koniono's sisters reportedly died. In Kpondu, Sierra Leone, Menindor was seeing more patients with the illness, and many of them were women and girls. She tried all her medical arts, all her secret weapons, but nothing seemed to work.

IDENTIFICATION

KENEMA GOVERNMENT HOSPITAL, SIERRA LEONE
March 13, 2014

Kenema is a small city, with a population of about 180,000, which sits at the feet of the Kambui Hills in eastern Sierra Leone. The Kambui Hills, long whalebacks covered with tropical forest, rise to the west of the city and look down on a maze of metal roofs and dirt streets. The countryside around Kenema is a hilly expanse of green that is dotted with villages and small towns. The communities are nestled among cassava and rice fields and patches of forest, and groves of oil palms, and stretches of thick farmbush. The soil is tawny orange and not terribly fertile, and the land is dissected by sandy streams and swamps. The sand contains diamonds. These days, freelance diamond miners can be seen working along just about every stream and watercourse around Kenema, sifting sand and mud through sieves and searching it for diamonds. Recently a Christian pastor who was sifting a stream to the north of Kenema found a yellow diamond the size of a lemon. Kenema is about a hundred miles by road from Kpondu, Sierra Leone, the village where Menindor the healer lived. The drive from Kenema to Kpondu takes five to six hours, and most of it is over rugged dirt roads.

The catchment hospital for the region is Kenema Government Hospital, a sprawling collection of low pavilions, linked by covered walkways and dirt roads, surrounded by a high wall, in the center of Kenema. Toward the middle of March, while Menindor was trying to heal sick patients in her village, a medical doctor and virologist at Kenema Government Hospital named Humarr S. Khan began receiving reports that a hemorrhagic fever had broken out in Guinea near the border with Sierra Leone.

Humarr Khan was then the chief physician of the Lassa Fever Research Program at Kenema Government Hospital. He was a virologist who specialized in Lassa hemorrhagic fever, which is a devastating and frequently fatal disease caused by Lassa virus, a Biosafety Level 4 virus that invades the brain and causes hemorrhagic bleeding. BSL-4 viruses, or Level 4 viruses, are sometimes referred to as hot agents. They are lethal, highly infective viruses for which, in almost all cases, there is no vaccine, no cure, and no effective treatment. If you catch a Level 4 virus, there is very little any doctor can do for you other than keep you hydrated and prevent you from having any contact with anyone else. In many countries, including in the United States, regulations require researchers who are handling Level 4 viruses to wear a whole-body pressurized biohazard space suit with an independent, filtered air supply. In addition, the space-suit research must be conducted inside a Biosafety Level 4 laboratory. A Level 4 lab is sometimes called a hot lab, a hot suite, or a hot zone. It is a group of rooms that are sealed off from the outside world and are only accessible through an airlock that is equipped with steel doors and a chemical shower. The shower is used for decontaminating the space suits of researchers when they make an exit from the hot zone.

The Lassa research program at Kenema Government Hospital is dedicated to giving care to Lassa patients, to tracing and stopping outbreaks of Lassa fever in the countryside, and to pursuing research on the virus in the hope of someday eliminating the disease. The program consists of a small high-biocontainment hospital unit, called the Lassa Fever Isolation Ward, a tiny group of offices, and a building called the Lassa Laboratory, inside which is a sealed, high-biocontainment laboratory known as the Hot Lab.

In the Hot Lab, the staff and scientists of the Lassa program wear high-biosecurity personal protection equipment, or PPE. A complete set of PPE gear amounts to a kind of nonpressurized space suit. It consists of a whole-body protective suit made of Tyvek (an impermeable fabric); a high-efficiency breathing mask called a HEPA mask, which can filter virus particles from the air; a transparent face shield or goggles; rubber gloves; and heavy rubber boots.

When Humarr Khan heard about a hemorrhagic disease breaking out in Guinea, he immediately suspected it was Lassa hemorrhagic fever. In addition to being a Biosafety Level 4 hot agent, Lassa virus is classified as an emerging virus. That is, the virus is making cross-species jumps out of nature into humans, and it is currently expanding its geographic range. The virus exists naturally in a certain kind of wild rat that lives in parts of West Africa. People come into contact with the rats, and the virus then jumps from a rat into a person. Once it gets into a person, Lassa virus can travel directly from person to person through contact with blood and body fluids.

Lassa attacks the brain and major organs, and it causes a polymorphic disease—one that takes different forms in different people. Some people who catch Lassa get a headache that lasts for about two weeks, and then the virus goes away—the person recovers fully.

Other people break with Lassa hemorrhagic fever. In these cases, the virus destroys the brain and causes failure of major organs. Lassa patients can have hemorrhagic nosebleeds, as well as bleeding from the mouth, eyes, and kidneys. They develop a swollen face; their hair can fall out; they develop a deadened, masklike facial expression; they can have seizures; they can fall into a coma that ends with an irreversible breathing arrest.

Lassa virus is thought to infect around 300,000 people a year in West Africa. It's not clear how many people die of Lassa hemorrhagic fever each year, but the fatalities certainly number in the thousands. Many of them are pregnant mothers and their unborn children. The Lassa program in Kenema is funded by the government of Sierra Leone and a consortium of research institutions around the world, including Tulane University, Harvard, and the Scripps Research Institute. Humarr Khan had many friends and colleagues at these institutions.

When he learned that a viral hemorrhagic fever was spreading in Guinea, he got in touch with some of his international colleagues to let them know about the outbreak, and he told them he would keep them posted on any developments. Khan was anticipating the arrival of many Lassa patients in his ward. He was seeing, he suspected, a big emergence of the virus from the wild rats.

BRUSSELS, LYON, GENEVA
March 13–21

In Brussels, Belgium, in a brick building on the Rue Dupré, managers at Médecins Sans Frontières, or Doctors Without Borders, a large international medical relief organization, were also receiving worrisome reports that a viral hemorrhagic fever had broken out in Guinea. The Brussels office of Doctors Without Borders is an operational center of the organization. The managers there quickly arranged for an investigation in Guinea. On March 13, a combined team of epidemiologists from Doctors Without Borders and the government of Guinea set out from Conakry, the capital of Guinea, in four-wheel-drive vehicles, to see what was going on in the Triangle. By this time, two and a half months had passed since the death of the little boy in Meliandou.

The team visited the hospital in Guéckédou, met with local health officials, and identified people with the illness. They also collected samples of patients' blood. The blood was sent by air to the Jean Mérieux-Inserm Laboratory, in Lyon, France, and to the Bernhard Nocht Institute for Tropical Medicine, in Hamburg, Germany. The French lab in Lyon is a Biosafety Level 4 facility, where scientists wearing biohazard space suits conduct research on Level 4 hot agents. There, a virologist named Delphine Pannetier, along with colleagues who included a virologist at the Pasteur Institute in Paris named Sylvain Baize, began working to identify the infectious agent in the blood samples from Guinea.

By the early hours of the morning on March 21, the French scientists were sure that the agent in the blood was a filovirus. The filoviruses are a family of viruses that all look alike, and most of them are extremely lethal. Sylvain Baize immediately sent an email to the head-quarters of the World Health Organization, in Geneva, Switzerland, announcing it was a filovirus. But, so far, the French and German teams couldn't tell what *kind* of a filovirus it was. They continued working at a frantic pace, trying to nail down the identity of the virus that was emerging in West Africa. The French investigators had learned that some of the patients had had hiccups. This detail really caught their attention. Hiccups are a classic sign of Ebola virus disease.

LISA HENSLEY

Eight hours after the French scientists identified the agent in the blood samples from West Africa as a filovirus, a virologist named Lisa Hensley was in an upstairs bedroom in her house, in Frederick, Maryland, pedaling a stationary bike. It was early in the morning, still dark. An icy, raw morning. As she pedaled, she took up her phone and began paging through emails and tapping out replies. At 5:48 a.m., an email from a colleague caught her attention: He was asking if she knew anything about an unidentified hemorrhagic fever virus in Guinea.

She typed while she pedaled: "Hearing it is Ebo. Don't know more yet—Info is confidential."

She finished her workout, showered, and got dressed. Spring was supposed to be on its way, but this morning was pure winter. She put on tights, a skirt, a sweater, loafers with block heels. Lisa Hensley has greenish-brown eyes and chiseled features, and brown, straight hair, which is lightened and cut so that it falls in curves along her cheekbones. She put on a small amount of makeup, and decided to wear her silver hoop earrings.

It was time to wake up her son. She went into his bedroom and looked up toward the ceiling. James slept on a platform bed above his desk. "Time to get up, sweetie."

There was a stirring near the ceiling, and eventually James's face appeared over the edge of the bed, looking down at her, his hair more moplike than usual. She guessed that he'd stayed up late surfing the Internet or playing games on his laptop.

James climbed down the ladder and started putting on his clothes, while she went back into her room. After a little while, she heard: "Mom, carry me downstairs." She went back into his room, and he jumped up on her back, and she piggybacked him down to the kitchen.

James had hemophilia, an inherited disease that causes the blood to fail to clot normally. A hemophiliac can bleed endlessly from a small cut, and can have dangerous internal bleeding after getting a blow to the head or body. James's hemophilia was mild and easily treatable, but when he had been a baby Hensley had decided to carry him up and down stairs to prevent him from taking a fall on the stairs, especially when he was learning to walk. Now, at age nine, he was an active, healthy, athletic boy. He went up and down the stairs all the time, except that she still carried him downstairs in the morning and upstairs to bed at night. It had become sort of a tradition, and they both liked it.

She dropped him off at school, and then drove to Fort Detrick. Cactoctin Mountain was a gray shape rising behind the Army base. The trees on the mountain were still bare, and under the rainy sky they looked more like smoke clinging to the mountain than trees. She drove through a security gate onto the base and parked next to the National Interagency Biodefense Campus. The campus consists of a group of buildings near the center of Fort Detrick. All but one of the buildings on the Biodefense Campus are starkly new.

Hensley went through a checkpoint into the campus and walked toward an L-shaped building called the Integrated Research Facility, or IRF. The IRF had just been completed, after nine years of construction. The facility is part of the National Institute of Allergy and Infectious Diseases, which in turn is a part of the National Institutes of Health, or NIH. The IRF's mission is to develop experimental drugs and vaccines, called medical countermeasures, that could defeat lethal emerging viruses and advanced biological weapons. Lisa Hensley had recently been appointed an associate director of the IRF. She was in charge of all of the scientific research programs at the facility. It was her job to manage all the IRF's research into the world's most dangerous viruses. She was new on the job, having started working at the IRF just two months earlier.

One wing of the IRF is faced with glass and resembles a fish tank, while the other wing is a brick-faced monolith, without too many windows in it, known as a biocontainment block. The block contains Biosafety Level 4 laboratories, where researchers investigate the most dangerous viruses on the planet. The IRF is the most advanced Biosafety Level 4 research facility in the world, and it is a crown jewel of the National Institutes of Health.

Hensley went through a second checkpoint and into the IRF, and she walked down a corridor to her office. Her office had a new-carpet smell. She'd taped to the door a colorful drawing of a giraffe done by James. She sat down at her desk and began reviewing the day's tasks. Meetings. Staff organization. Research programs, how to structure them, what to investigate. Budgets. Lab safety. The IRF's Level 4 labs hadn't yet gone into use—the space-suit labs were still awaiting safety certification by federal inspectors. That is, the IRF was still a cold facility, and it hadn't yet gone hot. The IRF would only go hot after its Level 4 labs had passed all their safety inspections. At that point, small frozen vials of Level 4 pathogens would be brought into the facility and placed in ultra-low-temperature superfreezers inside the IRF's Level 4 labs. The labs would then be hot, and the IRF would be a "supermax" prison for some of the fiercest life forms in nature.

As she worked on her email, Hensley couldn't stop thinking about Ebola. For sixteen years, she had been trying to find a cure for Ebola virus disease. There was no cure. No vaccine, either. No drug, no treatment, no nothing. Every time Ebola broke out it thrust doctors back into the Middle Ages. The only way to cut off an Ebola outbreak was to put people in quarantine camps, where they died like flies, as if they were in a fourteenth-century plague house. About the best doctors could do for Ebola patients was to give them water and hope for the best.

When Lisa Hensley was a junior in college, studying public health at Johns Hopkins University, she began thinking about the human immunodeficiency virus, HIV. The most recent evidence suggests that the most common type of HIV made a cross-species jump out of one chimpanzee into one person around 1910 in southeastern Cameroon, along a tributary of the Congo River. From that first human host, HIV began spreading from person to person, amplifying itself in the human species until it reached every community on earth. As of this writing, some seventy million people have been infected with HIV and thirty-five million have died of AIDS. As she describes it, Hensley had a sudden flash of conviction, as a college student, that emerging viruses were going to become the single greatest threat to human health in her lifetime. Right then, she decided to become a scientist and to search for ways to stop them just as they were coming out of the ecosystem, before one of them scored a hit on the human species. "Imagine if we had gotten ahead of the curve with HIV," she says, now. "Imagine if we had blood testing in place soon after it was discovered. Countless lives could have been saved. I wanted to do something about the next HIV."

In 1996, after getting a master's degree in public health and a PhD in molecular biology, Hensley got a job at the United States Army Medical Research Institute of Infectious Diseases, or USAMRIID (pronounced "you-SAM-rid"), at Fort Detrick. The main structure at USAMRIID is a vast, nearly windowless biocontainment block, built in the late 1960s and filled with a warren of hot zones. Today, USAMRIID is part of the National Interagency Biodefense Campus at Fort Detrick; it is the oldest facility on the campus. When she started working at USAMRIID, Hensley had no knowledge of space suits and no interest in working in a hot zone; she planned to do research on a mild virus that causes common colds, especially in children. This kind of cold virus infects many kinds of wild animals. Hensley thought that one of the wild cold viruses could jump out of an animal into a person somewhere on earth and start a global outbreak of a fatal, emerging cold.

The common cold seemed exciting to Lisa Hensley, but it wasn't very exciting to a colonel at USAMRIID named Nancy Jaax, who is an expert on Ebola virus. One day, after Hensley had been working at USAMRIID for about a year, Jaax asked Hensley if she could talk with her privately for a moment. "Nancy pulled me into her office and said, 'You're going to work with Ebola now,'" Hensley says. Hensley sensed that Colonel Jaax wasn't really offering her a choice.

Nancy Jaax and other Ebola researchers trained Hensley in the protocols of Level 4: putting on and taking off a pressurized biohazard space suit, passing through airlocks, and using a decon shower to sterilize the outside of the suit. Situational awareness—knowing exactly where your hands are at every moment. Handling needles and sharp instruments with *extreme* care. Hensley loved working in Level 4. Doing research in hot labs became almost an addiction for her, and she became fascinated with Ebola virus. It was really something to hold a flask containing a liquid suspension of ten billion particles of Ebola in her gloved hands inches from her space suit's clear faceplate visor. You always had to wonder what this virus could do to the human species if it got a big chance to amplify itself in human bodies.

Most medical investigators have zero interest in doing space-suit research on Level 4 viruses. Work in a biohazard space suit is physically exhausting and dangerous. It requires total concentration, and simple tasks take much longer to accomplish. And if you get a pinhole in your space suit and don't notice, a few particles of a Level 4 virus could slip inside and get friendly with you, and you wouldn't know what had happened until you started throwing up blood.

Lisa Hensley did research on virtually every experimental drug and vaccine that might conceivably stop or slow down an Ebola infection in the human body. She focused her work on a species of the virus named Zaire Ebola. Zaire was the first Ebola to be discovered, when it broke out in 1976 at Yambuku Catholic Mission in Yambuku, Zaire.

At this writing there are six known species of Ebola. The six Ebola sisters. In order of discovery, the six Ebolas are named Zaire Ebola, Sudan Ebola, Reston Ebola, Taï Forest Ebola, Bundibugyo Ebola, and Bombali Ebola. Each species of Ebola has its own identifiable genetic code and is somewhat different from the others. Zaire is the most lethal of the six Ebolas; it is the homicidal elder sister. In the 1976 outbreak, Zaire Ebola killed 88 percent of its victims, though in subsequent outbreaks it killed roughly 60 to 70 percent. Zaire Ebola is not only the most deadly of the five Ebolas, it is also the most deadly of all the known filoviruses, the family of viruses that includes the Ebolas. Zaire Ebola is the lord of the strains.

Lisa Hensley branched out into research on other emerging viruses. Her targets included Sars virus and Mers virus, which are lethal Biosafety Level 3 agents that circulate in animals and can invade people—and in fact these viruses *are* the animal cold viruses that Hensley feared, and they *do* cause extremely lethal cold-like illnesses in humans. Sars and Mers are contagious and can mutate very fast. Hensley studied Level 4 hot agents: Hendra virus, Lassa virus, monkeypox, Lake Victoria Marburg filovirus, and the Ravn filovirus. Ravn virus was first isolated from the blood of a ten-year-old Danish boy known as Peter Cardinal, who died of Ravn disease after possibly getting infected inside Kitum Cave, a bat cave on the slopes of Mount Elgon in Kenya. The Ravn virus has been found exactly nowhere except in the blood of the Danish boy—he is the only individual who is known to have been infected with Ravn. Still, Ravn could break into the human species again from its hidden reservoir in nature. Hensley also did research on a Level 4 emerging virus called Nipah. Nipah is a bat virus that causes personality changes and liquefaction of the brain. Nipah is only moderately infectious, but it gets into the lungs, and there is a certain alarm among experts that the code of the virus could change and turn the virus into a sort of brain-destructive neurological cough that travels in the air. There is no vaccine or treatment for Nipah disease.

Hensley also got involved with the smallpox virus. Smallpox, which was arguably the worst disease in human history, was declared eradicated in 1979. However, stocks of smallpox virus may exist in clandestine military laboratories of certain nations, including Russia and North Korea. Poxviruses such as smallpox are among the easiest viruses to modify with genetic engineering. Hensley and her colleagues had done research toward drugs that might protect the population from a genetically engineered superpox. Hensley had published 110 scientific papers, most of them involving medical countermeasures to emerging lethal viruses. Eventually she became the chief of research into antiviral drugs and vaccines at USAMRIID. Although she had risen to prominence, at least in the small, largely unseen world of people who work on defenses against the most dangerous elements of the virosphere, Hensley had no illusions about her importance as a scientist or a person. Biomedical research is done by teams. The research is time-consuming and deeply expensive, and the results are often disappointing. With persistence, talent, and luck, and plenty of money, a biomedical research team can sometimes pull the veil off some small mystery of nature and the human body, and can find a better way to treat a disease.

As the years went by, Ebola virus remained Lisa Hensley's oldest, most seductive enemy. She dreamed of finding a medical weapon that would stop the virus's growing entanglement with the human species. Angels used their swords to kill demons. In effect, the Ebola scientists hoped to find an angel's sword that would vanquish Ebola, slay it—a drug that could pierce Ebola's half-dead heart and make it die a permanent death. So far, there was no such thing as an angel's sword to kill Ebola. Now, in her office at the Integrated Research Facility, Hensley decided to wait until there was an official confirmation. She turned her attention to paperwork that morning, but she couldn't stop thinking about the fact that people were dying with hiccups.

INTEGRATED RESEARCH FACILITY
9 a.m., March 23, 2014

The French and German teams soon nailed down the exact identity of the filovirus in West Africa. On March 23, the World Health Organization announced that it was, indeed, Ebola: "A rapidly evolving outbreak of Ebola virus disease (EVD) in forested areas of southeastern Guinea. As of 22 March, 2014, a total of 49 cases including 29 deaths (case fatality ratio: 59%) had been reported."

After Lisa Hensley arrived at work and caught up on a little bit of email, she went down the hallway to the office of the director of the IRF, a virologist named Peter B. Jahrling. Jahrling is an expert in space-suit research, is a specialist in Lassa virus, and is the co-discoverer of Reston Ebola. He has a seamed face, a brush of gray hair, and looks as you'd expect a scientist might. He wears metal-rimmed eyeglasses, and often a grayish jacket with a subdued necktie in lighter tones of blue. Lisa Hensley had worked for Peter Jahrling for sixteen years at USAMRIID: They knew each other well.

Jahrling turned away from his computer. "Hey Lisa. What's up?"

"So it's Ebola," she said.

"Well, yeah. It was pretty surprising."

"Shouldn't we be involved, sir?"

Jahrling gave her an uncomfortable look. "How should we be involved?"

The IRF could put together a field team, she suggested. An away team. We could send the team to West Africa, she said to Jahrling, and try to save some lives. "I'll go myself, sir," she said. Meaning she would lead the team.

"I'm not in favor of you going to West Africa right now," Jahrling answered. The Integrated Research Facility was a *lab*. Staffed by . . . well . . . lab types. Young. Ambitious. Deft with pipettes and tiny vials of liquid. Stereotypable as nerds, though actually they were scientific hotshots. But to send an away team to Africa to fight a filovirus was a job for the Centers for Disease Control, not the IRF. And the IRF was in the middle of startup operations. Jahrling did not want his director of research suddenly flying off to Africa, taking staff with her.

Jahrling obviously had a strong point. Hensley couldn't really argue with him—objectively he was right. And she had James to think about. She was a single mother, and if she went to Africa to help people deal with Ebola, she wouldn't be with her son. One issue was his hemophilia. It was mild but a little unpredictable. One day he would take a tumble and he'd be fine, then another day he'd scrape his knee on the playground and it wouldn't stop oozing. James seemed to enjoy freaking out his teachers with a drippy cut. Hensley would get a panicky call from his school saying he'd cut himself and needed to be picked up. She'd take him home and watch his cut, and sometimes it would heal and sometimes it wouldn't. If it was still oozing, she would drive him to Johns Hopkins Hospital in Baltimore, where his doctors would give him clotting factor, and he'd heal immediately. The trips to Johns Hopkins were infrequent but unpredictable. As his mother, she felt she should be there for him. On the other hand, people were dying, and she was one of the few people in the world who knew much about filoviruses like Ebola.

Hensley went back to her office, feeling slightly less than completely useful. It had been more than a year since she'd put on a space suit and gotten her hands on a Level 4 virus. Her talent and accomplishments in the lab had led to a promotion into management. She had wanted to manage large research projects, of course. Now she did meetings all day, and collected a much bigger government salary. And still there was no vaccine or treatment for any filovirus, including Ebola. Nor was there a vaccine or medical countermeasure against other viruses that seemed ready for a breakthrough into the human species: Sars, Mers, Nipah. She wanted to go straight up against an emerging virus at the gates of the ecosystem, just as it jumped out of the virosphere and began to move into humans. She wanted to help people, try to save some lives, if she could. And the fact was, she missed her space suit.

The news came as a surprise to Ebola experts. Ebola virus had never been seen before in this region of West Africa. And it was Zaire Ebola, the hottest of the six Ebolas. It was the Ebola that had visited Yambuku Catholic Mission in late 1976. The Makona Triangle was more than two thousand miles from Yambuku, where the virus had emerged, killed some people, and vanished. Thirty-seven years later, Zaire Ebola had come out of nowhere in West Africa and was sacking humans along the Makona River. It was the lord of the strains, back from the dead.

RED ZONE

BRUSSELS, BELGIUM
CONAKRY, GUINEA
March 23, 2014

At the time of the WHO announcement, managers at the Brussels operational center of Doctors Without Borders had already started to go into action, organizing for another fight with Ebola. Over the years, Doctors Without Borders had become the shock cavalry of the human response to Ebola whenever it broke out. Ebola had to be put down quickly before it spread far and claimed many lives. Now, Doctors Without Borders began rushing medical supplies to Conakry, the capital of Guinea, and organizing teams of staffers and volunteers to go to the Makona Triangle and start putting down the virus.

Within days, people from Doctors Without Borders were setting up Ebola treatment units in Guéckédou and Macenta, the two small cities into which Ebola had now moved. A typical Ebola treatment unit of Doctors Without Borders is a collection of white plastic tents where Ebola patients are placed in strict biocontainment isolation so they can't infect other people. Patients are placed in cots in the tents, and tents are set up at the center of the camp, in an area known as the red zone. The red zone is surrounded by a maze of plastic fences, to separate infected people from everybody else. As long as a patient's blood tests are positive for Ebola, the patient isn't allowed to leave the red zone. Patients die in the red zone; they are not permitted to die anywhere else. When staffers exit the red zone, workers spray them with bleach before they take off their equipment, to sterilize it and kill any Ebola particles that are clinging to the gear. An Ebola patient who recovers is discharged from the red zone and allowed to go home. The bodies of patients who die in the red zone are placed inside double body bags, and are buried close to the camp. The red zone has pit toilets that are inside plastic sheds. There is a laboratory tent, where blood is tested, and there are generators to supply electricity.

The red zones of Doctors Without Border were, in effect, giant plastic bags in which people infected with emergent Ebola were kept. This technique trapped the virus inside the plastic bag, where it would work its way through the human bodies in the bag, killing many of them—but the virus couldn't escape from the bag. The red zones were artificial walls placed around hot spots of Ebola in order to break the growing chains of infection in the human species.

Each time Ebola started spreading in humans, Doctors Without Borders moved in with teams and tents and snuffed out the virus. The teams of Doctors Without Borders were very much like forest firefighters jumping into hot spots and putting down blazes while they are small. In the years since Ebola's first appearance, in 1976, there had been nineteen outbreaks of Ebola, and very few people had caught the virus. Never more than 280 people had died in any Ebola outbreak. In thirty-seven years of outbreaks, the six different Ebolas had killed, in all, only 1,539 people, according to reported deaths. The death toll from Ebola was virtually nothing in the annals of infectious disease: Each year, tuberculosis kills around 1.3 million people. Over the years, as Doctors Without Borders slammed down Ebola successfully, a widespread view developed among public health experts that Ebola wasn't much of a problem for the human population of the world and never would be. It is fair to say, however, that nature often does whatever is necessary in order to make the most number of experts wrong.

THE DOCTOR

The morning after the World Health Organization announced that Ebola had emerged in West Africa, Dr. Humarr Khan, the head of the Lassa research program at Kenema Government Hospital, in the city of Kenema, Sierra Leone, got up before dawn, as usual. Khan lived in a rented house on Sombo Street in downtown Kenema. That morning, he put on dark slacks and a short-sleeved shirt. He stuffed a wad of paper currency into his pocket, and he performed the dawn prayer on his prayer rug.

After praying, he went into the parlor. The room was dark and the curtains were closed. The room had a tiled floor, and it held a few pieces of furniture and a flat-screen television.

"Morning, Doctor," his houseman, Peter Kaima, said.

"Morning-o."

Kaima fixed a cup of instant coffee for Khan and handed it to him. While Khan drank the coffee, Kaima took a chicken sandwich out of the refrigerator for Khan. Khan put on a white baseball cap and slipped the sandwich into his office bag. Then he went outdoors to a courtyard, where an ambulance was waiting. It was a white 4x4 Toyota Land Cruiser with a diesel engine and rugged tires, a type of vehicle known in Africa as a bush ambulance, because it can go places nobody would believe. Khan climbed onto the front seat and chatted with the driver as the ambulance went down Sombo Street and turned onto Combema Road, a wide, dusty thoroughfare lined with small shops. Kenema is a maze of dirt streets and metal roofs. Rush hour was just starting, and people were walking along the sides of the road and were riding on bicycles and motorbikes, heading for jobs all over the city or heading out of the city to work in fields. The Kambui Hills, soft whalebacks covered with tropical forest, were beginning to catch a glow of first light. A smell of smoke from cooking fires drifted in the air, mixed with motorbike fumes and dust. This was the dry season of the year.

The ambulance went through the gates of Kenema Government Hospital. It is a sprawl of one-story stucco buildings surrounded by a high wall. The buildings are painted yellow and brown, or white and blue, and are linked by covered outdoor walkways. Dirt roads wander around the grounds, and clumps of flowering mango trees are scattered here and there, their thick crowns providing pockets of shade.

The ambulance dropped Humarr Khan by the adult wards, a group of low buildings near the center of the grounds. He went into the wards and began his morning rounds. The wards were large, open rooms that contained many beds lined up in rows. Nurses wearing pale blue uniforms were working in the rooms, tending patients and supervising their care. It was common for patients' family members to work alongside the nurses, giving care to their loved ones. Khan examined patients, spoke with family members, and gave instructions to the nurses. He also took time to train the nurses on various aspects of medicine and patient care, and he encouraged them to ask questions. "If there is anything you have a doubt about," he often said to the nurses, "I am ready to clear your doubts."

Humarr Khan was thirty-nine years old, a handsome man, not very tall, with a square face and a lively, intense manner. Khan's eyes were large and sensitive-looking, and set deeply in his face, and were cowled with strong eyelids, which could give him a veiled look. He was often enthusiastic and outgoing, but he could be secretive, too. He was single—divorced—and had a girlfriend whom he never seemed to talk about. His white baseball cap was a kind of trademark. Another trademark of Dr. Humarr Khan was an old white Mercedes sedan fitted with chrome spinner hubcaps. When he drove his Mercedes around Kenema, wearing his white cap, there were few in town who didn't know who that was. Khan loved soccer and was a passionate fan of A. C. Milan, the Italian soccer club. Some of Khan's American friends called him Cee-baby, a nickname taken from A. C. Milan.

As Humarr Khan did rounds in the general wards that morning, the hospital came alive. People walked in through the hospital gate or came in on motorbikes or in taxis. The hospital's walkways and porticoes got crowded with the families of patients. There were always children crying, people waiting anxiously outside a ward for news of a loved one, people resting in the shade of the mango trees, bush ambulances raising dust on the dirt roads as they bumped slowly past the wards and pavilions. Food vendors and drink sellers circulated along the walkways, holding trays of sandwiches and bottles of soda, keeping their voices low for the sake of the patients.

After he'd finished making rounds in the general wards, Khan walked across a dirt parking lot to his outpatient office. It was a white metal shipping container shaded by a roof made of woven palm fronds. The container had two windows and a door, but no air conditioner. Khan's waiting room was a set of outdoor benches next to the shipping container, shaded by a palm-frond roof. A number of walk-in patients were sitting on the benches waiting to see him. Many of them had arrived before dawn.

Inside Khan's shipping-container office there was a desk, a swivel chair, and a small exam table. Khan's walk-ins presented with dysentery, worms, skin sores, unexplained fevers, rashes, bleeding ulcer of the stomach, liver flukes, bacterial infections, spinal meningitis, heart failure, HIV, AIDS, and cancer. Patients with serious conditions often went to herbalists and faith healers first, and by the time they got to Khan's office it could be too late. He saw women with breast cancer whose tumors had broken through the skin, and he saw men with prostate cancer that had spread into their spines, leaving them paralyzed. He did what he could. For advanced cancer patients he prescribed palliative care. If a cancer patient could afford the cost, he'd send the person to Freetown, the capital of Sierra Leone, for treatment.

If a patient was thin or looked hungry, Khan would dig into his pocket and peel a few bills off his ever-present wad of currency and tell the patient to go buy some food. "You must eat or you cannot get well," he said to them. He also gave money to his patients so they could buy the medicines he prescribed for them. He took the money out of his salary and out of earnings he got from a private clinic he was running in town. A course of life-saving antibiotics could cost as much as twenty-five dollars. Not everybody in Kenema had immediate access to twenty-five dollars to save their life.

While Khan was rounding the general wards, a woman named Mbalu S. Fonnie was doing rounds in the hot zone of Lassa Fever Isolation Ward, a small white building that sat next to Khan's cargo container. Mbalu Fonnie was the supervisor of the Lassa ward, and she was an internationally respected expert on the clinical care of hemorrhagic Lassa patients in a high-biocontainment hospital ward. At the moment, she was wearing a cotton surgical gown, rubber boots, a surgical cap, double surgical gloves, eye protection, and a HEPA mask—a high-efficiency breathing mask that can stop a virus particle from getting into your lungs. Fonnie was a small, roundish woman in her late fifties, a very quiet person, deeply serious, a Christian, and she almost never smiled or laughed. She had once almost died of Lassa hemorrhagic fever. Having been carried to the edge of death by the virus, Fonnie thought she might have gotten *some* resistance to it now—but it's impossible to become fully immune to Lassa virus. She had been running both the Lassa ward and the hospital maternity ward for twenty-five years. Many younger people in town had been born under her watch in the maternity ward, and some of them had been born into her hands. Most people called her Auntie Mbalu or simply Auntie.

The hot zone of the Lassa ward consisted of a single, narrow corridor lined with nine small cubicles that opened on either side of the corridor. The patients lay on beds in the cubicles. The hot zone had a normal capacity of twelve patients—some of the cubicles held two beds, placed close together and almost filling the cubicle. There was a washing station with running water, where nurses could wash blood, feces, or vomitus off their gloves. At one end of the corridor there was an overflow room, and there was also a private nook, which was out of sight of the other beds in the ward.

On this morning there were only two patients in the ward, both suffering from Lassa fever. Two nurses were tending them. They wore the same kind of surgical outfit that Auntie wore. Like Auntie, all the nurses in the Lassa ward were survivors of Lassa fever, and so they were thought to have a bit of resistance to the virus.

Auntie examined the two patients. Then she walked to the end of the corridor, to the exit door of the hot zone. She opened the door and stepped outdoors, into fresh air, crossed a small space, and went inside a cargo container. The container was the dressing room and staging area for the hot zone. There, she removed her surgical gear. Underneath she was wearing a flawlessly clean, starched white nurse's outfit. She put on a small white nurse's cap, and went outdoors and around the corner to the main entrance of the Lassa ward, where she went into a foyer and sat down by the nurses' charting table and waited for Humarr Khan. Every morning they met in the foyer of the Lassa ward.

After seeing walk-ins, Khan went next door to the Lassa ward and found Auntie sitting at the table. This morning he had big news. Yesterday the World Health Organization had announced that the disease in Guinea wasn't Lassa fever, as Khan had first suspected, but was Ebola hemorrhagic fever. The disease was similar to Lassa but had a much higher death rate, and Ebola virus was much more infective than Lassa virus. And Khan told Auntie that Ebola had a long record of killing medical workers. The Lassa ward was the only high-biocontainment medical unit in Sierra Leone. The staff was well-trained and had years of experience handling bleeding, infectious Lassa patients. Auntie Mbalu Fonnie and her nurses would be the front line if Ebola reached Sierra Leone.

Auntie was a person of very few words. She often spoke under her breath in a kind of whisper, and she had a British accent. She would have listened intently to Khan as he described Ebola to her, absorbing everything he said about the disease. His words were somber. When he had finished, she probably would have responded with something like, "Well, God will take control in this situation. God will drive it." She also may have said to him, "God holds in God," which was one of her favorite expressions. It means that God holds all the power and keeps his plans hidden until events come to pass.

After meeting with Auntie, Khan walked downhill along a dirt road to a construction site in a corner of the hospital grounds, where a cluster of unfinished buildings stood. The buildings, sophisticated structures made of concrete blocks, would become a new Lassa ward. Khan walked behind a cargo container and sat down on a plastic chair, out of sight of the rest of the hospital, and lit a cigarette. Khan never let hospital staff or patients see him smoking. He'd put the chair behind the cargo container in order to create a hidden smoking place. As he smoked, he thought about Ebola. In the next hours and days he would be speaking to the entire staff of the Kenema hospital, getting them informed about the virus. He planned to read up on Ebola and talk with colleagues who were studying the virus. He would also look into possible experimental drug treatments, in case there was any sort of drug that might help save an infected patient.

Humarr Khan had been running the Lassa program at the Kenema hospital for ten years. His predecessor had been a physician named Aniru Conteh. In 2004, a pregnant woman with Lassa hemorrhagic fever had a bloody miscarriage in the Lassa ward. Afterward, she had a profuse hemorrhage from her birth canal and went into shock from blood loss. There were no blood supplies in the ward, so Dr. Conteh couldn't give her a transfusion. He decided to give her an intravenous infusion of saline solution—sterile salt water—to try to stabilize her. He placed the infusion needle a vein in her leg. After the infusion, he pulled the needle out of her leg. Then, acting from habit, he tried to put a plastic cap over the bloody needle in order to make it safe. The needle missed the cap, went through two layers of surgical glove, and pricked him lightly in the finger. Dr. Conteh hardly noticed the prick. Ten days later, he died of Lassa fever in his own ward, while Auntie Mbalu Fonnie and the Lassa nurses cared for him. They were crying behind their surgical masks as he passed.

Afterward, an American doctor named Daniel Bausch set out to hire a new director for the Kenema Lassa program. Dan Bausch, a professor at the Tulane School of Public Health and Tropical Medicine, in New Orleans, was the American liaison to the Lassa program, and he had been a close friend of Dr. Conteh. He flew to Sierra Leone and started talking with doctors in the capital, Freetown, trying to find somebody who'd agree to replace Dr. Conteh. "If you had canvassed doctors in Sierra Leone about their dream job," Dan Bausch said to me recently, "running the Lassa fever ward in Kenema would have been dead last for most of them." Kenema was a remote city in the diamond fields, the government salary was pitiful, and the Lassa ward was now an obvious death trap for its director.

After weeks of fruitless searching, Bausch ran across Humarr Khan. Khan was then twenty-nine and fresh out of his residency from the College of Medicine at the University of Sierra Leone. Bausch invited him to have a beer with him at a hotel in Freetown. After a brief chat, he offered Khan the job.

Khan didn't immediately accept. Bausch raised the stakes: He painted a picture of a possible future for Khan. Lassa was obviously a huge problem, and Khan would help save lives if he took the job. He would do research on Lassa virus with leading American scientists. He would speak at international conferences. He would likely become a co-author of scientific papers in top scientific journals. The government salary was terrible, Bausch added.

Khan asked for a day or two to think about it. What he actually needed to do was meet with his father. A young man had to get his father's consent for an important decision. His parents, Ibrahim and Aminata Khan, were living in a town by the sea on the far side of Freetown Bay, opposite the city of Freetown. Mr. Khan was 91, a retired educator with a national reputation, and he was a strict disciplinarian. Humarr was the youngest son among a total of ten brothers and sisters.

They thought of him as their baby brother, bright but irresponsible, and some of them still called him by his childhood nickname, Squazzo. Now, he rode a rusting ferry across the bay, and took a taxi along a dirt road to a shady neighborhood of small houses made of cement blocks. In the sea just off the beach, fishermen in long wooden boats were setting nets, and smoke from cooking fires drifted through the neighborhood, mixed with a salty tang of the Atlantic Ocean.

He sat down with his parents on the veranda, and spoke to them in Fulah, the family's ethnic language. He told them about Dan Bausch's job offer.

Mr. Khan practically blew up. "It will be dangerous to work with Lassa virus!" he exclaimed in Fulah. "Look what happened to Dr. Conteh." It had been all over the Freetown papers.

"Don't worry, sir. I know how to keep safe."

"You do *not* know how to keep safe!" Mr. Khan said heatedly.

Ms. Khan agreed. She didn't want her son going anywhere near Lassa virus.

"Don't do it," Mr. Khan added.

"But it's what I want to do," Humarr answered.

In their view, this was the whole problem with Squazzo: He did what he wanted. They thought he had gone off the rails while he was in medical school in Freetown. He drank beer, he smoked cigarettes, he partied late at night with friends, he hung out at bars and nightclubs, and he had girlfriends. "You are headed straight to hell! Straight to hell!" Mr. Khan had warned him. Now, he urged his son to forget about Lassa fever and move to America. "Young people are coming over to the U.S. and making a lot of money." Humarr's older brother Sahid was living in Philadelphia, where he was an IT specialist. "Sahid can help you get established in Philadelphia."

"I don't want to live in Philadelphia. I can't work in an office, sir. I have to go out and be a doctor."

"So be a doctor in Philadelphia. Or Baltimore."

"I won't go to America. I will stay here, sir," Humarr said. The next day, he told Dan Bausch he'd take the job.

A decade later, Dan Bausch's predictions for Humarr Khan had come true: He was doing research with top American scientists, and some of them had become his close friends. He spoke at international conferences. He had co-authored scientific papers in top journals, though he hadn't yet made it into *Science,* which would be a peak achievement in any scientist's career. The government salary was indeed terrible, but he had set up a private practice in Kenema that was bringing him some income. As he settled into work in Kenema, Khan was very aware of what had happened to his predecessor, Dr. Conteh. He did not often put on PPE and go inside the Lassa ward. One tiny accident in the hot area could cost you your life.

After finishing his cigarette, Khan came out from his hiding place behind the cargo container and walked along a dirt road to the Lassa program office. It is a small, one-story stucco building with a palm tree growing in front of it. A few ambulance drivers and staffers were usually gathered under the tree, sitting on a bench and chatting, waiting for an ambulance call. Khan said hello to them and went into the office of the Lassa program coordinator, a young woman named Simbirie Jalloh. He asked her if there were any emails or phone messages. She told him that one of his scientific collaborators, an American woman named Pardis Sabeti, was setting up a conference call about Ebola, and she hoped he could join the call.

SABETI

Over the years, Dr. Pardis Sabeti had forged ties with the Lassa program, and had become friends with Humarr Khan. When Khan joined her conference call, she was sitting at her desk in her office at her laboratory in the Northwest Building at Harvard University. Other scientists were speaking from other locations. Just then, Sabeti's pet brown rat, Coco, was either asleep in her lap or was exploring Sabeti's office, which the rat did fairly often. ("People probably think I'm psychotic, but I don't do caged animals," Sabeti says.) Pardis Sabeti was then an associate professor of biology at Harvard. She is a slender person, then in her late thirties, with a warm manner. She specializes in reading and analyzing the genomes of organisms. In addition to running a lab at Harvard, Sabeti leads viral genome efforts at the Broad Institute of MIT and Harvard. In particular, she studies virus evolution—the way viruses change over time as they adapt to their environments. In her spare time, she is the lead singer and songwriter for an indie band called Thousand Days. The band's fourth album ended up getting delayed due to the Ebola outbreak.

"Humarr, how are you?" Sabeti asked. "I'm worried about you. I'm worried Ebola could get to Sierra Leone."

Khan said he was worried, too. The Lassa program's Hot Lab was the only high-biocontainment laboratory anywhere across an extended region of West Africa. The first principle of warfare against an emerging virus is to know where it is moving. Yet Humarr Khan had no lab machines that could identify Ebola virus in human blood. If Ebola crossed the river and entered Sierra Leone, Khan and his team would need to be able to identify people who were carrying Ebola. If infected people could be identified, then they could be isolated inside the Lassa ward and cared for by specially trained nurses wearing hazmat gear. This would stop the virus from getting into other people, and it would break the chains of infection.

Sabeti offered Khan a special device known as a PCR machine. It could detect the genetic code of Ebola in human blood, and so it could be used to test patients' blood. She said she'd send him one immediately, along with people who could train Khan's Hot Lab staff in how to use it.

After talking with Khan, Sabeti went out of her office, shutting the door behind her so her rat wouldn't get loose, and she drove to the Broad Institute, which occupies a pair of crystalline buildings in Kendall Square next to the MIT campus. About four thousand scientists work regularly at the Broad, where they decode and analyze the genomes of organisms. At her offices on the sixth floor, Sabeti called a meeting with a few people, a group that would grow in size and would be known as the Ebola War Room Team. They began planning and directing elements of the human defense against Ebola. Sabeti raided her Harvard lab budget for $600,000 worth of equipment, lab supplies, and cash. By the end of the day, she had given two of her colleagues, Kristian Andersen and Stephen Gire, the assignment of taking the stuff to Kenema and setting up an Ebola blood lab for Humarr Khan. Andersen and Gire made preparations to depart for West Africa as soon as possible.

INTEGRATED RESEARCH FACILITY
FREDERICK, MARYLAND
Next day, March 25

Humarr Khan had been reading up on Ebola, and it alarmed him. The day after he spoke with Pardis Sabeti, he talked with a scientist named

Joseph Fair, who was then working for an American biotech firm called Metabiota. Khan and Fair were close friends and drinking buddies, and Khan had once saved Joseph Fair's life. Fair offered to set up a blood lab for Khan in Kenema, and Khan said sure, absolutely. Pardis Sabeti was sending him a blood lab, too, but he wanted to play it safe.

As soon as he had made plans with Khan, Joseph Fair reached out to his friend Lisa Hensley at the IRF. He asked her if she'd like to come with him to Kenema and help set up the lab for Khan. Hensley knew Khan pretty well. She had gone to Kenema with Fair once before, and the two scientists had worked together setting up a Lassa blood-testing lab for Khan.

Lisa Hensley liked Humarr Khan, and the idea of helping in the Ebola outbreak still really intrigued her. She went back to her boss, Peter Jahrling, and asked him if she could get a short leave of absence from the IRF so that she could travel to Kenema and help Khan. "I can do this," she said to Jahrling.

"I couldn't argue with Lisa this time," Jahrling now says. He knew Humarr Khan, too, and liked him. He contacted his boss at the NIH, and they worked out a deal: The NIH would *loan* Lisa Hensley to the U.S. Department of Defense for a three-week military deployment to Kenema Government Hospital. Hensley was a civilian, but while she was in Africa she would be operating inside the U.S. military chain of command.

Hensley soon received a letter from Humarr Khan:

Dear Dr. Hensley,

It is with great pleasure I write to invite you to help with your expertise on epidemic preparedness and outbreak response on Ebola fever. . . . The very long border we are sharing with Guinea increase[s] the susceptibility of a possible import of this disease to our country.

Meanwhile, Pardis Sabeti's scientists had already arrived in Kenema, and they quickly finished setting up their blood-testing equipment in the Lassa program's Hot Lab.

At this point, the U.S. military changed Hensley's deployment orders. She had to go where the military sent her, and she was assigned to Monrovia, the capital of Liberia, to set up an Ebola blood-testing lab for the government of Liberia. Several cases of Ebola had been reported in that country. She wouldn't be going to Kenema; she wouldn't be working with Humarr Khan after all.

Hensley and Joseph Fair went into the supply storerooms of the IRF, as well as the storerooms at USAMRIID, and they began assembling all the elements of a portable space-suit lab for testing blood. They began packing everything into military trunks for air shipment to Liberia. One evening, when Hensley was at home and putting her personal things into a duffle bag, James offered to help her pack. He went into her closet and got out a wide-brimmed beach hat and stuffed it into her duffel bag. "You need a hat so you won't get sunburned in Africa," he said.

It was just around then that Menindor the healer, the sage with almond eyes who kept a magic snake in her house, fell sick in Kpondu village. As Hensley packed for Africa, Menindor lay in bed in her house in Kpondu, with vomiting and diarrhea, while her elderly mother and her sister cared for her. For the many people who depended on Menindor and cherished her, her illness must have been a frightening thing to see. It seemed that not even Menindor's powers could overcome the evil force of the sickness that was visiting the villages of Sierra Leone.

SURVEILLANCE

A hundred miles from Menindor's village, Dr. Humarr Khan, at Kenema Government Hospital, had seen no reports of Ebola in Sierra Leone. The virus was active in Guinea, just across the Makona River. He worried that the virus could cross the river, traveling inside a person, and could start spreading through villages in Sierra Leone. The Lassa program had a team of epidemiologists, called a surveillance team. The team traveled in bush ambulances, for carrying suspected Lassa patients back to the Kenema hospital, where they would be placed in the Lassa ward if they tested positive for Lassa.

Khan met with the surveillance team in a room called the Library, which is in the Lassa Laboratory. There weren't many books in the Library, but it had a couple of tables and an Internet connection that sometimes worked. Khan told the team that it was going to be hard to distinguish Ebola disease from Lassa fever. Both viruses produce similar symptoms—diarrhea, vomiting, extreme pain, bleeding from the orifices, shock or coma, death. If the surveillance team found anybody showing Ebola-like signs, the person was to be fitted inside a hazmat suit, to protect the team members from infection, and taken by ambulance back to the Kenema hospital, where the patient's blood would be tested for Ebola as well as for Lassa.

The surveillance team visited villages along the Makona River, asking questions, describing the symptoms of Ebola, and looking for anybody who might have it. They came up empty-handed. The villagers said they hadn't seen this disease.

The Kenema surveillance team was being advised by an American scientist named Lina M. Moses. A lively woman in her thirties, Moses was a postdoctoral researcher at the Tulane School of Public Health and Tropical Medicine, in New Orleans. She was an expert in infectious disease ecology, which is the study of the way ecosystems, viruses, and people interact. Each year, Moses lived in Kenema for several months, doing research. She spoke Krio, one of the two national languages of Sierra Leone (the other one is English). She also spoke some Mende, an ethnic language. When Moses was stationed in Kenema, she spent time in the countryside, trapping rats and testing their blood for Lassa virus, watching the virus's hidden movements in the ecosystem and through people.

In 2011, a molecular biologist named Erica Saphire visited the Kenema hospital, where she met Lina Moses for the first time. "My impression was of a sweaty, dirty young woman, with dark hair and an expressive face," Saphire recalls. "She was wearing a work shirt, jeans, and hiking boots, and she was carrying a plastic bucket full of dead rats. The rats were floating in a liquid that looked like the beginnings of a Cajun gumbo," Saphire says. Lina Moses's rat gumbo was hot enough to frighten Army scientists at Fort Detrick. Saphire had immediately liked Moses.

Moses began meeting frequently with the surveillance team in the Library. She gathered their reports and made recommendations for where they should go next. She thought about where Ebola could be hiding in Sierra Leone, if, in fact, it had crossed the river. She spread out maps of eastern Sierra Leone on the table in the Library and studied them. Moses knew the territory fairly well.

But her knowledge of the countryside was partly in her head. The maps of the Makona Triangle aren't reliable. The landscape is dotted with tiny villages, typically spaced a few hundred yards apart. Different villages often have similar or identical names. In Krio the same word can often be spelled several different ways. Therefore the name of a village could appear under several different spellings on different maps. Furthermore, many villages were just a dot on the map—no name at all. And there were uncharted villages, small communities that didn't appear on any map. Some of them were reachable only on footpaths.

Lina Moses made calls to community health clinics in and near the Makona Triangle. She told the clinic managers about Ebola and asked if they'd seen any patients with the symptoms. She also called district medical officers and asked them to track down reports of suspicious illnesses. The Lassa surveillance team visited villages, gave the local people information about Ebola, and asked them if they'd heard of anybody with an illness like it. The surveillance team and Lina Moses came up empty-handed. The local people were saying that they hadn't seen anything like Ebola disease.

GUINEA
Last days of March

On the Guinea side of the river, staffers and health workers with Doctors Without Borders were identifying Ebola patients and placing them in the red zones of tented camps, where the patients were given care according to the rules of Doctors Without Borders. Those who died in the red zone were buried near the camp. As the virus continued to spread, local people got increasingly fearful of the Doctors' treatment units. The camps, with their white tents, had a sinister look, and they were being run by white foreigners. The foreigners were telling people that they were infected with a virus and must go into the camp, and those people were vanishing in the camps, never to be seen again. Foreigners dressed in moon suits were carrying white bags with dead bodies inside them out of the camps and burying them next to the camp. Nobody was allowed to open the bag and look inside at the loved one, because the foreigners said the corpse was dangerous.

People in the Makona Triangle had never heard of Ebola. Many people didn't believe in the infectious theory of disease—the idea that diseases are caused by invisible microbes. Plenty of people in Kissi country had mobile phones, and they used social media to discuss the situation. Text messages started flying around the Makona Triangle, rumors about the camps. The white foreigners, went the rumors, were injecting chlorine into people and doing hideous experiments on their bodies.

A doctor named Armand Sprecher, who is a public health expert at Doctors Without Borders, later explained to me that the Doctors' Ebola treatment centers always awakened fear in local populations. "In every Ebola outbreak we were seen as strange people in strange clothes carrying out experiments on poor Africans, or harvesting organs," Sprecher said. "We were supposed to be profiting in some way. People have thrown plenty of stones at us."

If a group of powerful foreigners who spoke no English, or spoke it badly with a heavy accent, were to set up a camp of tents in suburban Wellesley, Massachusetts, and they were wearing biohazard moon suits and were telling townspeople that an extreme virus had gotten loose in Wellesley and that anybody who had symptoms must go into the camp and stay there until they died, there might be some opposition from the Wellesleyites. And if most people who went into the camp were never seen again, dead or alive, and if the foreigners were burying white body bags next to the camp, and if quite a few of the bags obviously held dead children, and if social media lit up with rumors of hideous experiments, it's a pretty sure bet that the Wellesleyites would be reaching for their guns and doing anything they could to get the hell out of Wellesley. "It's your darkest nightmare," Armand Sprecher explained.

By March 31—just a week after the WHO announcement that Ebola had broken out in Guinea—the number of reported cases in Guinea had jumped from 49 to 112, with 70 reported deaths. The fatality rate was holding steady at around 60 percent. People in the Makona Triangle were getting nervous. Many didn't believe in this thing called Ebola, but they were sick. Some of them went to other villages, to get care from family members or treatment from healers, and some headed for cities, where they thought they might get better medical care than in a tent.

Doctors Without Borders was beginning to get nervous, too. On March 31, an official at Doctors Without Borders sent out a news release that vibrated with alarm: "We are facing an epidemic of a magnitude never before seen in terms of the distribution of cases in the country."

Officials at the World Health Organization reacted to the Doctors' report with skepticism. A WHO spokesman named Gregory Härtl tweeted a response, saying that the outbreak was "relatively small still" and there was "no need to overblow something which is already bad enough," and "there has never been an Ebola outbreak larger than a couple of hundred cases." And this new outbreak was no bigger than any of the others—it was just a typical Ebola outbreak.

Lina Moses was watching the situation from Kenema. She had never worked in an Ebola outbreak, nor had Humarr Khan. They met in the Library and talked about the situation. They both agreed that Ebola was likely to reach Sierra Leone, if it hadn't done so already. It was just common sense to think that the virus would cross the river. At the same time, Moses and Khan felt that there wouldn't be more than a few dozen Ebola cases in Sierra Leone, at most. There was no reason to suppose that Ebola would go far in Sierra Leone; the consensus view among public health experts was that Ebola was not a serious threat to the human population of the earth, and that the virus could be controlled fairly easily.

KPONDU VILLAGE, SIERRA LEONE
April 1–8

Menindor the healer was now very sick, lying in bed in her house. It was a small house made of mud bricks. Menindor's mother, who lived in a nearby village called Sokoma, moved into Menindor's house and took care of her, but there was nothing she could do to help her daughter. Menindor's sister also cared for her. On April 1, while Menindor lay dying in Kpondu, a doctor in Freetown named Jacob Maikere, who was working for Doctors Without Borders, circulated by email an English translation of a report from Guinea's Ministry of Health. Maikere sent the email to Sierra Leone's Ministry of Health and to Lina Moses.

The report had come out on March 24, a week earlier, and had originally been in French, the language of Guinea. The report listed some Ebola cases that had supposedly occurred in Sierra Leone or in Liberia. One of the cases was that of Ms. Sia Wanda Koniono. She was the woman from Kpondu who had gotten sick after traveling in Guinea and riding in a taxi or in a jitney bus and sitting next to a sick person. The report said she'd died in the Guéckédou hospital in Guinea on March 3. It said she came from a village in Sierra Leone called Peluan, that she had traveled to Guinea, and that she had died there and been buried near a village called Gbandu. The report didn't say where "Gbandu" is. In fact, Gbandu is Kpondu, in Sierra Leone—the village where Menindor the healer lived.

Lina Moses was having trouble getting her email. The Internet connection in Kenema was bad, and so she didn't see the email in question. The report seemed to say that a woman got Ebola in Guinea, died in Guinea, and was buried in Guinea. It stated that the woman's dead body was taken to "Gbandu," but the report did *not* say that Gbandu is in Sierra Leone. In fact, there is a village in *Guinea* called Gbandou. Gbandou, Guinea, is three and a half miles from Kpondu, Sierra Leone. And the name of Menindor's village in Sierra Leone has four different spellings, *Kpondu, Gbandu, Koipondu,* and *Koipind*. How could anybody guess that Gbandu and Koipind are the same village, and are in Sierra Leone? This illustrates the nightmare of identifying villages in the Makona Triangle.

Menindor died in her house in Kpondu on April 8. Her death rocked the villages along the Makona River. As the news spread that Menindor was gone, there was an outpouring of grief, especially among women. Menindor's family began planning a large funeral, and hundreds of people from villages on both sides of the river made plans to attend. Outside the villages, Menindor's death went completely unnoticed.

FLASH

KPONDU, SIERRA LEONE
April 8–10, 2014

Menindor's sister and their mother had been caring for Menindor in life, and now they cared for her in death. After they had washed the body internally with an enema, they washed the outside of it with water. As they poured water over the body, the women may have collected the dirty water in a container such as a plastic laundry tub. Maybe they didn't save the wash water they poured over the body. We simply don't know.

I was in Kenema doing research for this book during the time when the crisis was receding, and one evening I was drinking a beer with a Sierra Leonian public health expert named Macmond Kallon, and he told me something I hadn't heard before. Kallon had been working for the World Health Organization, visiting villages in the countryside around Menindor's village, searching for people with Ebola and trying to get them into treatment centers to keep them from spreading the disease to others.

"The whole secret of the transmission of Ebola," Kallon said, "is that you wash the body with water, and then the water is collected and used again." The wash water from the corpse was carefully stored in a container. Then family members would use the water in a ceremony of grief and remembrance. "If you are the son of the dead person, you wash yourself with this same water, the water that washed the body," Kallon explained. "After that, the daughter washes herself with the same water that the son used." The washing ceremony was sometimes done in a grove of sacred trees, a patch of old forest. It was a type of ceremony similar to the one in which mourners ate a meal of the deceased person's favorite foods while sitting next to the body of the deceased person in a forest. During the water ceremony, family members sometimes drank the wash water, Kallon said. It was a way of bringing the essence of the deceased person into themselves.

Ebola victims sweat profusely. The sweat glands pour out vast numbers of Ebola particles mixed with perspiration. The sweat clings to the skin and evaporates, leaving a film of Ebola particles behind. More sweat comes out, and more particles build up on the skin. By the time a person dies of Ebola disease, the corpse is painted with Ebola particles. There can easily be ten million Ebola particles sitting on a square inch of the corpse's skin. Only one particle can cause an infection in another person. Ebola particles are quite tough as long as they are kept moist. Experiments have revealed that Ebola particles can stay potent for up to seven days while they're sitting on the skin of a decaying corpse.

Thursday, April 10

Menindor's funeral was held two days after she died. At least two hundred people showed up for the funeral, most of them women and young girls, and their grief was dramatic. Menindor lay on a bier, wrapped in fine cloth, her face and perhaps her arms and hands exposed to view. The mourners wept over her body, caressed her face, embraced her. As people touched the body, Ebola particles on the body's skin were transferred to the mourners' skin and clothing, especially to their hands. The mourners touched and hugged one another during the funeral, and they wiped tears from their eyes with their fingers. It was a huge crowd, moving around the bier of Menindor, expressive, churning, torn with loss.

I remember when my father died. I was with him at the time. My mother was there, too. As his breathing settled down, my mother held him close, and I put my arms around my mother. Then, after a few moments, I reached out and placed my hand on my father's face, feeling his skin, still warm to the touch, my father, as the life went out of him. I couldn't help but touch him as he went away. At the funeral of Menindor, as people expressed their grief and touched Menindor and touched one another, some of the particles that had once coated her bare skin were transferred from person to person until the crowd got smeared with Ebola. People got particles on their fingers and hands, on their faces, in their hair and clothing, and in their eyes. Ebola virus moves from one person to the next by following the deepest and most personal ties of love, care, and duty that join people to one another and most clearly define us as human. The virus exploits the best parts of human nature as a means of travel from one person to the next. In this sense the virus is a true monster.

If a single, well-formed particle of Ebola lands on the wet membrane of a person's eyelid, it can slip through the membrane in seconds, entering a tiny blood vessel. From there the particle is pulled into the system of veins that lead toward the heart. The particle is a tiny thread of flotsam, jiggly and rubbery, flipping and flopping and twisting in the currents of the blood. It bumps against red blood cells but slides off them, doesn't stick to them. If the particle were the size of a small shred of spaghetti, then a red blood cell would be the size of a dinner plate. The Ebola particle passes through the heart and lungs, traveling in the flow of the blood, and it enters the system of arteries, which lead away from the heart to all parts of the body. Sixty seconds after the Ebola particle has landed on the person's eyelid it can be anywhere in the person's body.

Eventually, somewhere in the person's body, the particle sticks to a cell. The core of the particle gets pulled inside the cell. Now *one* Ebola particle is sitting inside *one* cell in the person's body. At this point, the person may be doomed.

Inside the cell, the core of the Ebola particle falls apart. Its RNA, its genetic code, comes spooling out of the particle's broken core, like thread whirling loose off a spindle. Next, the code takes command of the cell's machinery and forces the cell to start making copies of the Ebola particle. Eighteen hours later, the cell is oozing newborn threads of Ebola, which grow out of the cell like hair, break off, and are carried away in the bloodstream. Each infected cell spits out up to ten thousand new Ebola particles. The particles end up everywhere in the body, infecting more cells, and each cell spits out thousands more Ebolas. This is known as extreme amplification of a virus. Soon the body is flooded with particles, and the person's immune system collapses. At the time of death, vast numbers of cells all through the person's body are converting themselves into Ebola particles. Ebola particles are made *entirely* out of human material: Ebola is an anti-human metamorphosis of the human form. The amplification of Ebola virus in the human body is one of the dark wonders of nature.

After Menindor's funeral in Kpondu Village, the mourners returned to their villages elsewhere in the Triangle, where eventually some of them got sick. Those who loved them gave them care, and the virus moved to the caregivers, traveling along chains of duty and affection. The Ebola parasite got into a human network of affection and care, the ties of humanity that join each person ultimately to every other person on earth.

Some of the infected individuals sought help elsewhere, help from doctors and hospitals, from family members, from healers, and in different nations within the Makona Triangle or outside the Triangle. The network of human connection, going hot with a virus, extended into the cities of West Africa. The funeral of Menindor appears to have been a central, germinal event in what would become a full-scale epidemic of Ebola in the human species, the most destructive, fast-moving expansion of any lethal infectious agent during the past hundred years.

When epidemiologists finally learned about Menindor's funeral and followed the chains of infection that emerged from the funeral, they found that at least 365 cases of Ebola could be traced back to the funeral. Menindor's funeral set off chains of infection running in all directions, into Liberia, into Guinea, and toward Kenema Government Hospital, seventy miles away. The chain reaction had begun when a small boy touched an animal, possibly a bat, and a few particles of Ebola crossed the uncertain boundary that separates the body of a human being from the rest of the natural world.

Seven weeks after the parasite jumped out of nature and amplified itself in the boy, it entered Ms. Sia Wanda Koniono of Kpondu village, perhaps when she was riding in a jitney bus and bumped up against a sick passenger. She died on March 3 and was buried near Kpondu. Twenty-eight days later, on the first of April, the email sent by Jacob Maikere reporting Ms. Koniono's death and burial went unnoticed by Humarr Khan and his team.

If Khan's team had seen the report and had realized its significance—that Ebola was already in Sierra Leone, and was active in Kpondu, then it seems certain that they would have sent a surveillance team to Kpondu village to find out what was going on. The team would have arrived in Kpondu shortly after April 1. They likely would have found Menindor dying of Ebola in her bedroom.

If the Kenema team had discovered Menindor, they might have been able to get her into isolation, thus protecting people from exposure to Ebola. The Kenema team also might have been able to prevent a large, public funeral for Menindor from taking place. If Menindor's funeral hadn't happened, would Ebola virus have expanded into the world in racing chains of infection? Would the outbreak have gone so far and so deep? Would Ebola have reached Dallas, Lagos, and New York? If Menindor had been found in time, a hot seed of the epidemic might not have sprouted, and the outbreak might have been more local and more manageable.

Maybe or maybe not. We will never know, because this is not the way reality played out. And this is not to say that Humarr Khan and his team are at fault for anything. They are not at fault. It is to say that history turns on unnoticed things. Small, hidden events can have ripple effects, and the ripples can grow. A child touches a bat . . . a woman riding on a bus bumps against someone who isn't feeling well . . . an email gets buried . . . a patient isn't found . . . and suddenly the future arrives.

Ebola is not a thing but a swarm. Ever since a few particles of Ebola had slipped into the boy, the virus had been copying itself in ever larger numbers of people. It was a vast population of particles, different from one another, each particle competing with the others for a chance to get inside a cell and copy itself. As the particles copied themselves, there were errors in copying, and slightly different kinds of Ebola emerged in the swarm. You could imagine the virus as a school of fish, with each particle of Ebola a fish. The fish were swimming, and as they swam and multiplied they changed, until the school had many kinds of fish in it and was growing rapidly in size, with some kinds of fish better at swimming than others and with sharper teeth.

By the time Menindor caught the virus around March 26, the virus that had originally infected the little boy had mutated into several different kinds of Zaire Ebola. The virus had been chaining through people, crushing their immune systems, exploring the body's defenses, and it had begun to adapt to the human species. Sometime in early March, a new kind of Ebola arose in the Makona Triangle. This mutant Zaire Ebola was *four times* better at infecting human cells. The mutant had a special affinity for human cells. It was a fish with sharper teeth. Menindor got infected with the mutant, and it killed her. The mutant came out of her body in death and was spread through her funeral.

The change in Zaire Ebola happened in just one letter of its genetic code. The virus that was spread at Menindor's funeral, the mutant, the changed Ebola, is officially named the A82V Makona Variant of Zaire Ebola. In this book it will be referred to simply as the Makona strain or the hot Makona. Scientists think it first arose as a *single* mutant particle of Ebola in the body of some unidentified person who lived near or in the city of Guéckédou, Guinea. That one mutant particle of hot Makona, that one fish with sharper teeth, multiplied to vast numbers in the victim's body. Then, somehow, the hot Makona strain got to Menindor. She almost certainly caught it from one of her patients. As of this writing, however, many questions about the hot Makona remain unanswered.

The Makona strain really gets into human cells easily. The strain *may* be able to spread through a person's body faster and more powerfully than any other kind of Ebola—but this has not been proven. The Makona *may* be more contagious than any other Ebola, easier to catch. It could be hotter—more lethal—than any other filovirus, even hotter than Zaire Ebola. We really don't understand the exact characteristics of the Makona strain. But by now there is quite a bit of evidence that the Makona strain is the most contagious and devastating kind of Ebola that's yet appeared.

In the funeral of Menindor we are seeing something that resembles a high-speed movie displaying the first instants of a nuclear explosion. We are looking into the core of the expanding fireball right at the start of detonation. The funeral produced an unseen biological flash of a new virus, and the virus began explosive amplification in the human species.

Ninety miles away, at the Kenema hospital, nobody saw the flash, nobody was aware of it. The staff of the Kenema hospital didn't know that something had happened that would lead to the deaths of many of them. Nobody anywhere noticed the movement of the mutant virus as the Makona strain began to spread in chains of infection, chains that branched into more chains and more branches. The hot Makona had started chaining its way toward every human body on earth, a biological wildfire growing in the human species. While much remains unknown about the hot Makona, there is no doubt that by the first day of April it had become the new lord of the strains.

BREATHING UNIT

It was a beautiful spring day in eastern Maryland, the first sunny day in a long time, warm and windy, and puffy white clouds were marching across a blue sky. Lisa Hensley was due to leave for West Africa the next day, and she had spent this day packing biohazard gear into trunks. In the afternoon, she drove to James's school to pick him up. The road went over the Monocacy River, near the place where the Battle of Monocacy was fought during the Civil War. Oak trees growing on the bluffs along the river were covered with a haze of buds as red as wine. On the backseat of the car she'd left a battery-powered breathing unit for a lightweight portable space suit.

James lifted his wheelie bag, full of books, into the backseat, and climbed into the front seat. He had noticed the breathing unit. "What's that, Mom?"

"It's a PAPR," she said. "That means positive air pressure respirator. It runs from a battery, and I attach it to my suit. It filters the air I'm breathing inside the suit. And it keeps my suit pressurized."

"Are you going to be wearing your suit in Africa?"

"Yes, sweetie, I'll be wearing it."

"What would happen if I got Ebola, Mom?"

James's question took her by surprise. She saw immediately that he'd been looking up Ebola on the Internet, on his laptop. He had probably seen pictures of people with Ebola disease, too.

As for James's question, she had no idea what Ebola would do to him. Ebola had never been observed in a person with hemophilia. Ebola makes people bleed, and their blood won't form clots. The same thing happens with hemophilia.

After a slight pause she said, "Well, you'll be in a big bubble, right? The bubble's to keep you from infecting anybody else."

"So what will you be doing in Africa?"

She glanced at him. Was he worried about her? They were driving through a neighborhood where daffodils were blooming in people's yards behind chain-link fences. They passed an auto repair shop. "I'm going over to set up ways to help people. It's testing, so they can know if they've got Ebola."

"You're making a medicine for Ebola?" he asked.

"We don't have a medicine for Ebola yet."

"But don't you work on a vaccine?"

"Yes, sweetie. I do work on a vaccine. But we don't have a vaccine for Ebola."

James screwed up his face with a quizzical look. "Have you been doing this for a while?"

She smiled. "Since before you were born."

"Well, what *else* have you been doing, Mom?"

She almost burst out laughing. "What else? Well . . " Hmm. Writing scientific papers. Doing her best to be a good mother. Trying to keep herself and her team members safe in the hot lab. Staying close to her parents as they got older. Pushing forward with research on medical countermeasures to other emerging viruses, like Sars. Having a relationship with Rafe, her off-and-on boyfriend. But not marrying Rafe. She had long felt that her romantic life was somewhat screwed up. After her divorce from James's father, she had decided not to marry again, or maybe it was just that she hadn't actually found the kind of love that happens in fairy tales. "Lots of things, sweetie," she said.

Next morning, while James's school was starting, she took him to an IHOP and they ate too many pancakes. Then she dropped him off at school, really late, and gave him a good kiss, not too huggy or sentimental. Her parents would be looking after him—they'd driven to Frederick from their home in North Carolina.

Hensley parked her car at Fort Detrick and climbed into a government van that was full of military transport lockers packed tightly with all the elements of a rapid-deployment Biosafety Level 4 field lab for testing blood. The van took Hensley and her things to Dulles International Airport. That evening Hensley and her crates were in the air over the Atlantic Ocean, heading for Monrovia, the capital of Liberia.

THE SNAKE OF MENINDOR

MONROVIA, LIBERIA
Same day

Eternal Love Winning Africa Hospital occupies a tract of land, nearly a mile long, that lies along the beach just south of the city center of Monrovia, Liberia. ELWA Hospital is a Christian institution, and is staffed by Liberians and foreigners, many of whom are Americans. Quite a few of the foreigners stay and work at the hospital on temporary tours of duty. They live in bungalow houses scattered along the beach, and they work in a group of concrete hospital buildings that are clustered around a small chapel. The beach is lined with palm trees, the air is warm and salty, and a steady sound of Atlantic surf, punctuated with the cries of seagulls, drifts across the hospital grounds. An American evangelical medical relief organization called Samaritan's Purse has a medical mission at ELWA Hospital, with a small, two-story office building made of concrete blocks that have breezeway holes in them.

On the day the World Health Organization announced that Ebola had broken out in West Africa, the emergency operations director of Samaritan's Purse at ELWA Hospital, Dr. Lance Plyler, started making preparations, in case Ebola patients began arriving at ELWA Hospital. Lance Plyler is a slender, slight-figured man in his forties, with dark hair and an intense manner. Plyler asked one of his colleagues at Samaritan's Purse, an American doctor named Kent Brantly, to set up an Ebola ward at ELWA Hospital. Lance Plyler and Kent Brantly were close friends. They had worked in medical emergencies with Samaritan's Purse in various places around the world.

Kent Brantly began looking around the hospital grounds for a good place to set up an Ebola ward with a red zone for the biocontainment of patients. Eventually he decided to put the Ebola ward in the hospital's chapel. Brantly made the chapel into a red zone—he put up plastic barriers, and moved medical supplies and protective gear into a storeroom near the chapel. The ward had five beds for Ebola patients. After that, Kent Brantly, Lance Plyler, and the rest of the Samaritan's Purse staff watched and waited.

April 12

A Marine lieutenant colonel attached to the U.S. Embassy in Monrovia met Lisa Hensley at Roberts International Airport when she got off her flight. They loaded the transport chests into some Embassy vehicles, and the vehicles traveled a short distance along a dirt road out of the airport and into a tract of forest, and arrived at the Liberian Institute for Biomedical Research, or LIBR. The facility is a former chimpanzee research station that has been converted into the principal medical testing and research lab of the government of Liberia.

Hensley and her colleague Joseph Fair met the staff and explored the facility. It is a collection of concrete buildings, which includes large chimpanzee houses that have barred doors and windows. The chimp research had been stopped, and the chimpanzees had been moved to an island in a nearby river.

Several rooms at the former chimp station were now well-equipped laboratories for research on HIV. Hensley and Fair thought these rooms would be a good place for setting up a Hot Lab for testing blood samples for Ebola. They started unpacking their trunks, and they met a team of Liberian lab technicians who would work in the blood lab and would eventually take over the operation of the lab after Hensley had returned to the United States. That evening, an Embassy car took

Hensley and Fair into Monrovia and dropped them at a hotel near the beach. The hotel would be their home for the coming weeks.

Early the next morning, an Embassy car carried Hensley and Fair back to the former chimp station, and they began working as advisors to the team of Liberian government lab technicians, setting up the hot lab and showing the techs the equipment. The techs hadn't done high-biosafety space-suit work before. Then, right in the middle of this, Joseph Fair suddenly left the assignment and instead went to work as an advisor to the Ministry of Health of Sierra Leone.

Hensley continued working with the Liberian team, and together they created a hot lab. They sealed the windows and doors of a suite of rooms, creating an air-controlled lab. They installed negative-pressure HEPA air circulators, which lowered the air pressure inside the lab and filtered the air coming out. The filters would block any particles of Ebola or microscopic droplets of blood that might be in the air inside the lab, and keep them from escaping. They created a decontamination room, a gray room between the hot lab and the rest of the building, where they would spray their suits with bleach before removing them, to kill any Ebola on the outside of the suit.

When they had finished the preparations, the Liberians and Hensley turned on the HEPA circulators and lowered the pressure in the lab, and then tested it to make sure the seals were good. The lab was now ready to go hot.

Hensley showed her colleagues how to put on and take off a whole-body pressurized lightweight space suit. The suit had a battery-operated respirator attached to the waist of the suit—a PAPR. For a helmet, the suit had a clear, soft plastic bubble that surrounded the person's head and provided a complete field of view. By April 15, three days after Hensley had arrived, the team had the LIBR Ebola lab at the former chimp station up and running, and they immediately began testing samples of human blood for the presence of Ebola virus. The samples were sent to the lab by doctors from all over Liberia.

A few miles away from the former chimp station, doctors at ELWA Hospital, on the beach in Monrovia, were collecting samples of blood from patients and sending them to Hensley and the LIBR team. None of the blood that arrived at the LIBR lab came up positive for Ebola. It seemed to Hensley and the other observers that Ebola had gotten into a few people in Liberia and then had vanished. Hensley's lab, so far, had not actually gone hot. It had not yet had Ebola virus in it.

By the day Hensley arrived in Liberia, there had been ninety-four cases of Ebola reported in the country, with forty deaths. Seven of those cases had been in Monrovia, a city with a population of a little over a million. The virus had seemed poised to take off in Monrovia, but then the virus began to weaken in Liberia. In the last week of April, there were only thirteen new cases of Ebola reported in the whole country. The outbreak seemed to be fading.

We now know that the virus in Liberia at that time was genetically close to the Zaire Ebola that jumped into the little boy in Meliandou. In other words, the first kind of Ebola to enter Liberia was wild Ebola, fresh from the ecosystem. The A82V Makona Variant, the Makona strain, the mutant, which had first appeared in Guinea just a few weeks earlier, was still jumping around the Triangle, and hadn't made a breakout for the cities. And nobody knew the Makona strain existed.

Hensley worked twelve-hour days in the lab. She found her Liberian colleagues to be dedicated professionals, very good at Level 4 lab operations. She didn't feel disappointed by the fact they weren't finding Ebola in the blood the doctors were sending to the lab, but it seemed as though she had missed a chance to get her hands on Ebola virus and actually fight it. Her hotel on the beach seemed pretty luxurious, though she didn't have time to swim in the ocean. Every night before she went to sleep she called home using Skype. She told James she'd been wearing her sun hat, the one he'd packed in her duffle bag for her. She thought he seemed unworried.

James was worried. One evening Hensley's parents told her that he had had a hard time at school that day. His class had been studying the life of the Russian composer Tchaikovsky, who had died of cholera. The disease causes fatal diarrhea with sudden death, and James had gotten quite upset when he learned about the symptoms. Finally the teacher took him out of the room and asked him what was wrong. He had said that he was worried about his mother. Eventually the teacher brought him back into the classroom, and he told the class that his mother was in Africa helping people fight Ebola, a fatal disease that was like cholera. "I'm so proud of my mother," he had said.

On April 28, she flew back to the United States, having never found Ebola in any blood sample. Her Liberian colleagues continued blood testing operations in the Liberian national lab. Hensley returned to work at the Integrated Research Facility. She continued getting the research programs organized and started. She stayed in touch with her Liberian colleagues at the former chimp station. As the weeks passed, they saw no Ebola in the blood samples that were coming into the lab. Ebola seemed almost gone.

MAKONA TRIANGLE
Later April

The sister of Menindor the healer, the sister who'd given care to Menindor and had washed her body, died in Kpondu some weeks after Menindor died. Menindor's mother, who'd also given Menindor care, fell sick in her village, and died at some time in April or possibly in early May. On April 21, Menindor's husband died. Nine days later, Menindor's grandson died in Kpondu. By early May, the virus was hitting women in Buedu, Nyumundu, Kolosu, Fokoma, and Sasani. These women had been at Menindor's funeral or were closely connected to women who'd been at the funeral. All of them died of the hot Makona strain.

The death of Menindor's husband impressed people and frightened them. Many people in the area believed that diseases were caused by malignant forces. They were aware of the story that Menindor had been keeping a supernatural snake inside a chest in her house. After she died, a story went around that Menindor's husband had very foolishly opened the chest. The reptile had escaped and killed him, and now the snake of Menindor was going around the villages, planting its invisible fangs in people, destroying them. The villagers had their own explanation for what was happening.

Even as the Makona strain was quietly amplifying itself in the villages, the number of new reported Ebola cases in West Africa continued to drop. Zaire Ebola, the Ebola that had gotten into the little boy in Meliandou, was starting to wane in the human species, even as the Makona strain was growing new chains of infection. One fish was pulling ahead of the others.

In the hot lab at the Liberian national lab in the former chimp research station near the airport in Monrovia, the Liberian government technicians continued testing hospital blood samples for Ebola—but Liberia seemed to be free of the virus. In Kenema, Humarr Khan and Lina Moses continued to be on the alert for Ebola cases, and they continued to send surveillance teams out into the villages looking for signs of virus, but so far not a single case had been detected in Sierra Leone. With Ebola seeming to fade away, Humarr Khan decided to attend a scientific conference in Daytona Beach, Florida. He left Kenema on April 25.

AMERICA

After the conference in Florida, Humarr Khan flew to Baltimore to visit with two of his sisters, Isatu and Umu, and a brother, Alpha. Alpha was working in sales in Baltimore, and he knew the city well. Maybe a little too well. Humarr stayed out late with his brother and did some bar hopping around the city, and they generally enjoyed themselves.

Humarr's brother Sahid, the IT manager in Philadelphia, was sixteen years older than Humarr, and he had taken on the role of a parent for his younger brother when Humarr was growing up in Sierra Leone. Humarr still addressed Sahid as "sir," as he had done ever since he was a boy, and he also called Sahid by the nickname "Brother J."

Sahid began to get a little bit worried about Humarr, running around Baltimore with Alpha. Eventually he phoned Humarr for a talk. "Have you guys been doing all that stuff Dad says will send you straight to hell?" Sahid asked. (Their father was now ninety-nine and in good health.)

"Not so bad, sir."

"Well, what did you do? You and Alpha?"

"We only had a couple of beers to have fun," Humarr answered.

"What about marijuana? Did you smoke marijuana?"

"No, sir. Never. I don't smoke marijuana. I smoke cigarettes, sir. Alpha doesn't even smoke cigarettes."

Sahid laughed gently. "Okay, look, I have no problem with having a good time. But I always like to preach moderation with you."

"Yes, sir."

Sahid was getting amused by all the sirs coming from Squazzo. "Sir? Who are you calling sir?"

Humarr laughed. "Yes, Brother J."

Sahid had a feeling that his brother was vulnerable. For years, he had been worried that Humarr could die in the Lassa ward. And now there was Ebola to worry about. "America is safe from these things," he said to Humarr on the phone. "Just stay here. Don't go back to Sierra Leone. You can live near your sisters there in Baltimore."

"I don't want to live here, sir."

"But you can come to Philadelphia. You can stay with me. I can help you get set up here. You can be a doctor here."

"They don't need me in the cities of America," Humarr answered his brother. "They need me for the Lassa fever over there."

"Yes, but do you know how much doctors are making here? Six figures."

"But I'm a trained virologist, sir." He sounded completely confident in himself. "I'm very good at what I do."

Humarr had recently spent three years in Ghana getting advanced medical training. Sahid and their sister Umu had loaned him $17,000 to pay the tuition, since his paltry government salary couldn't cover the cost. Sahid reminded him that in the United States a doctor can earn $17,000 with relative ease.

"I have to pay for my training myself," Humarr said. Committed to returning their gift, he had started a private practice in Kenema, and the income from the practice would eventually help him pay back the money. He also told Sahid that he would be returning to the United

States in just a month—in June—to take a fellowship at Harvard with Pardis Sabeti.

"When you're in Cambridge, I hope we have more time to talk about your moving to the United States," Sahid said.

"Yes, sir. We'll talk."

Sahid ended the conversation feeling that he missed his younger brother badly. He really didn't like the idea of Humarr going back to Kenema.

DARK WINGS

May 7

After his visit to the United States, Khan returned to his duties at the hospital. The Ebola outbreak seemed to be fading. On May 7, health authorities in Conakry reported that the number of new Ebola cases in the city was dropping sharply: The city was seeing only about one new case a day. In Liberia, Ebola had disappeared. And not a single case of Ebola had been reported in Sierra Leone.

KPONDU VILLAGE
May 7

One of the leaders of the Kenema surveillance team was a man named Michael A. Gbakie. A soft-spoken man in his forties, with a quiet demeanor and a sensitive gaze, Gbakie (pronounced "Bocky") was the chief biosafety officer in the Lassa research program. He traveled in a white Land Cruiser with a driver and, typically, another surveillance officer. They stopped in villages and described the signs and symptoms of Ebola, and asked if anybody knew of or had seen anybody with this disease.

On May 7, Michael and his team learned that there had been some deaths in a tiny village, Kpondu, in an area called the Kissi Teng Chiefdom. They drove to Kpondu, and found a few small, rectangular houses with metal roofs arranged around an open space. Next to the houses stood cooking sheds roofed with palm fronds—in the tropical climate, people did their cooking outdoors. The team identified themselves as being from the Ministry of Health and said they were looking for a disease called Ebola. They described the signs and symptoms, and asked if any deaths had occurred in the village recently.

The residents said that a famous healer, a woman, had died in the village. She had been a leader of women. Her name was Menindor. She had become sick and died.

Michael and his partner asked the villagers about the cause of death of this healer: Did it seem like Ebola?

No, she had failed with the law of the smoke, the villagers said.

Michael had no idea what that was. He wasn't going to argue with the villagers about their explanations for diseases. He just wanted to know if the healer's disease had seemed like Ebola.

No, it wasn't Ebola, the villagers said. In addition to failing the law of the smoke, the healer had failed her responsibility with the snake, and this was the cause of her death.

Michael had absolutely no idea what they were talking about.

Menindor's husband had also died, they said. He had opened his wife's bag, or chest, and had seen the snake inside it. Her husband saw the snake and that was why he had died. And then a little boy had died. The reason for the boy's death was that Menindor had prepared a magic flagstone and placed it in the lintel of the door to her house. Menindor's magic stone had been removed from the doorway and this had caused the boy to die, they said.

Gbakie and his partner were using a translator who was helping them with the Kissi language. Using the translator, they said, "But we want to tell you that Ebola is close to you," they said. "You have to be very watchful for Ebola." If anybody had symptoms of this disease, they should go to the community health clinic in a town called Koindu. The head of the Koindu clinic was in contact with Humarr Khan, and would report any Ebola-like diseases to Khan if he saw them in his clinic.

Michael Gbakie and his team returned to Kenema on May 8, thinking that the death of the healer Menindor had occurred a short time before then. In fact, her death had occurred exactly a month earlier, on April 8. The Makona strain had been moving out of her funeral for a month, and had not yet been noticed. It seems that the villagers were aware of Ebola, but possibly they didn't want to disclose all they knew out of fear they would end up in a camp. And so Michael Gbakie and his team had come close to the Makona strain, but it had slipped past them.

By the third week of May, the outbreak seemed essentially over. During the entire time of the outbreak, there had been a total of 258 cases of diagnosed or probable Ebola in West Africa. Zaire Ebola was acting just the way it had acted in all previous outbreaks: a couple of hundred people get Ebola, and the virus goes away, suppressed by the Ebola fighters. The World Health Organization prepared to announce the end of the outbreak; the announcement was planned to come at the end of May. Doctors Without Borders made plans to start closing their Ebola treatment units and bringing their staff home.

The dry season was ending in West Africa, too. The rains arrive in Kenema during the last few days of May. As May 2014 came to a close, the sky remained blue and cloudless. But as the days went by, the hot, dry air grew a little strange, a little restless, got a prickly feeling. The sky was a smooth blue void, except for egrets flying, small white things moving high up. A local legend said that the egrets were watchers who looked down on the world and protected all the other animals on the earth: cattle, goats, crocodiles, lizards, even snakes. The protection of the egrets did not extend to the flies, however. The flies were too small and too numerous for the egrets to see them or keep track of them. Thus it was said that the flies had to look after themselves, and they were well able to do that. At night, when the city became quiet, Equatorial stars glittered in a black, cloudless sky, and heat lightning flashed in the southwest, a sign of approaching storms. After so many months of heat, of smoke from burning fields hanging in the air, and a lion-colored sky filled with dust blowing in from the Sahara Desert, people ached for rain. They wanted the world to be lashed with storms and turning emerald green; and they wanted to hear the roar of water pouring through the gutters that ran along the streets, washing all the grime out of the city.

KENEMA
May 20

As the rains approached and Ebola disappeared, Lina Moses, the American scientist who was advising the Kenema surveillance team, began feeling deeply homesick. She lived in a small house in New Orleans with her husband, an artist named Aron Belka, and their two young daughters. She had been stationed in Kenema for months, and missed them terribly. She found Khan in the program office and told him she would be leaving.

"I'd like you to stay longer," Khan said to her. He was still worried about Ebola, he said. The virus hadn't completely died out in Guinea, and it could still cross the river. If even a single case of Ebola popped up in Sierra Leone, there would have to be an intense search for the virus in the population. He really needed Lina Moses to stay in Kenema in case the virus appeared.

Moses couldn't stay. Her husband had been taking care of their daughters, driving to and from school and daycare, cooking, taking them to play dates, letting them play in his studio in the afternoons while he painted. She couldn't bear to be without her children any longer. She left Kenema and took a series of flights home. A day and a half later she embraced her husband and children, weeping profusely as she held them, incredibly happy to be home.

While Moses was in the air over the Atlantic Ocean, a twenty-year-old woman named Victoria Yilliah was admitted to the maternity ward of Kenema Government Hospital with a high fever and bleeding from her birth canal. She had been pregnant, and had gone into labor at the community health center in Koindu, in the Kissi Teng Chiefdom. Ms. Yilliah's baby had been born dead, and had come out in a flood of hemorrhagic blood. Her husband, Anthony, brought her to the Kenema hospital.

The staff of the maternity ward examined her. When they gave her an injection and set up an intravenous drip of saline for her, she had hemorrhages around the needle punctures in her skin—her blood had lost its ability to clot, and it ran steadily out of the needle holes. Her symptoms pointed to a fatal case of Lassa hemorrhagic fever. She was transferred to the Lassa ward, where she came under the care of head nurse Auntie Mbalu Fonnie.

RAIN

At the Lassa ward, Auntie put on PPE and examined Victoria Yilliah. It seemed almost certain that she would die. Lassa virus is especially lethal in women who are pregnant. The virus infects both the mother and her fetus, which dies and may be spontaneously aborted, while the mother has profuse hemorrhages from her birth canal and typically also dies. Nevertheless, Auntie had been able to save the lives of some infected mothers by giving them an abortion followed by a dilation and curettage, or D&C. A D&C is a procedure in which a curved scalpel called a curette is used to scrape out the inside walls of the uterus, removing any remaining fragments of the placenta. The Lassa rescue procedure was a risky, last-ditch effort to save the life of a woman who otherwise would probably die. If the fetus could be removed quickly, and a D&C performed afterward, the procedure seemed to offer the mother some chance of survival. The baby could never be saved.

Ms. Yilliah had already lost her baby. Auntie decided to give Ms. Yilliah a D&C in an effort to save her life. She gathered a team; all donned PPE. Fonnie scraped out the patient's uterus with a curette, assisted by her team, and the procedure went normally. Afterward,

Ms. Yilliah rested quietly in a bed in the Lassa ward, tended by Auntie and her nurses. She didn't die. Auntie wasn't especially surprised by this. Victoria Yilliah fought for her life for days in the Lassa ward, and eventually she began to improve. In the end, she would survive her illness. She would be discharged from the ward in June and go home to her husband, Anthony. The D&C procedure seemed to be a factor in her survival.

While Victoria Yilliah was fighting for her life in the Lassa ward, thunderstorms built over the Kambui Hills, piling up in white towers. Crackles of lightning ran from cloud to cloud, and thunder boomed, and a downpour fell over Kenema and pounded the hospital. The rains had arrived.

The thunderstorm didn't last long. Afterward the sun came out, and the metal roofs of the city steamed in the heat. But then another thunderstorm arrived. The rains came in waves, storm by storm, building up over the Kambui Hills, releasing sheets of rain. Lightning bolts started hitting the ground, and eventually the storms merged and produced steady downpours.

As the rains rumbled, Khan got a phone call from the medical director of the Koindu clinic, where Victoria Yilliah had had her hemorrhagic miscarriage. He said that a female patient in his clinic had the symptoms of Ebola. Furthermore, he said, two women with Ebola-like symptoms had been taken out of the clinic by their relatives and had been brought to the Kenema hospital.

This was really alarming. Khan asked the clinic director to send a sample of the sick woman's blood to Kenema to be tested for Ebola. And it seemed that there were two women somewhere in the Kenema hospital who might have Ebola. Khan phoned a staff doctor at the hospital named Abdul Azziz Jalloh and asked him to immediately search the wards for any female patients showing signs of Ebola.

Dr. Azziz rounded the wards, and discovered a woman who had classic signs of Ebola. "Her conjunctiva [mucous membranes of the eyelids] were inflamed, and she had a mask-like face, red eyes, diarrhea, vomiting, dry lips, and inflamed gums," Dr. Azziz recalled later. The woman's name was Satta K. He ordered a blood test and had her transferred to the Lassa ward. Satta K. ended up lying in a bed in the Lassa ward near Victoria Yilliah.

The next morning, a messenger riding a motorbike left the clinic in Koindu. His motorbike held a plastic box, and inside the box there was a single glass tube containing an amount of blood that would barely cover a person's fingernail. The blood had been collected from the woman with Ebola-like symptoms in the clinic; her name was Mamie Lebbie. She happened to be a sister-in-law of Menindor the healer, and she had been at Menindor's funeral. The motorbike messenger hit thunderstorms, and reached Kenema in the late afternoon, and delivered the blood tube to the director of the Hot Lab, a man named Augustine Goba.

Mr. Goba put on a white hazmat suit, goggles, double gloves, and rubber boots. Carrying the blood tube, he pushed open the door to the Hot Lab. A whisper of air flowed around him as he opened the door, traveling inward to the hot zone. He worked on the blood sample, preparing it for testing. This was a time-consuming process. It got late; he would continue the work the next day, which was a Sunday.

HOT LAB
KENEMA, SIERRA LEONE
Sunday, May 25

On Sunday morning, some of the nurses who came on duty at eight o'clock in the general wards sang hymns in order to lift the spirits of the patients and themselves as they began their day's work. On Sundays, Muslims in Kenema tuned in to Christian radio stations and listened to gospel choirs singing in three-part harmony, just as, on Fridays, Christians in Kenema listened to the sermons of imams on the radio. As the morning went on, Christian families poured into the streets, walking to church. People were dressed well, men in sport shirts and slacks, boys dressed like their fathers, young girls wearing white or pink dresses, and women wearing long, brightly colored skirts and head wraps that matched. Some people were texting or talking on cellphones as they walked along.

That morning, in a small lab inside a white cargo container in an alley next to the Lassa Laboratory, a French scientist named Nadia Wauquier (pronounced VOH-kee-ay) was preparing to join the effort to test the blood sample that had arrived from the clinic in the Makona Triangle. Nadia Wauquier was an employee of the American biotech firm Metabiota, and she was stationed in Kenema doing surveillance for emerging viruses. She had a PCR machine in her lab that could detect the genetic code of Ebola in human blood. For weeks, she had been using the machine to test samples of blood taken from patients at the Kenema hospital. She was monitoring the hospital's patient population for any appearance of Ebola.

Augustine Goba would run one test on the blood sample from the clinic in the Makona Triangle using a PCR machine provided by Pardis Sabeti at Harvard. Nadia Wauquier would run a separate test on the blood using her machine. With two tests, the results could be cross-checked.

Now, Nadia and her assistant, a female technician named Moinya Coomber, suited up in PPE and entered the Hot Lab. They began purifying a very small amount of the blood sample—a drop the size of a sesame seed—in order to test it. Augustine Goba arrived in the lab and began working with a separate droplet of the blood. The preparations took time, and there were seven other samples of human blood to prepare for testing as well—Nadia was continuing to do routine blood tests of patients in the hospital. Humarr Khan arrived and sat at a table in the Library, across the hall from the Hot Lab, waiting, wondering what was going on. Rain fell and stopped, and the sun came out.

5:30 p.m.

Nadia was in her cargo-container lab, holding a thin glass tube in her hands. It contained a purified extract of blood taken from Mamie Lebbie, the sick woman in the rural clinic. She fitted the tube into a tray of her PCR machine—a device about the size of a microwave oven. In the tray there were seven other thin glass tubes, each of which held a purified sample of blood taken from a patient at the Kenema hospital.

These seven samples were just routine; Nadia had been testing patients' blood for months, and not one sample had ever been positive for Ebola. She started the machine. It contained eight samples, in all.

An hour later, just as darkness came, she saw that her machine had almost finished its run. By then, Augustine Goba and Humarr Khan had arrived in her lab. They gathered around a computer screen, and the finished results came up.

As she looked at the results, Nadia could see that the test had gone wrong. Of the eight blood samples, *three* of them were positive for Ebola. One of the positives was for Mamie Lebbie. The other two positives were patients at the Kenema hospital. Three people with Ebola, two of them in the hospital?

"This can't be right," Nadia said.

They talked it over and decided these were false positives.

"Let's redo the test to be sure," Khan said. Then he went outdoors, to an alley next to the lab, and phoned the minister of health of Sierra Leone, a woman named Miatta Kargbo. He had been keeping in touch with her all day, letting her know that Ebola might have arrived in Sierra Leone. "Madam Minister, there have been some technical issues. We are going to do the test again."

The minister was really unhappy. "I want you to stay by the lab, Dr. Khan," she told him. "You have to be there right to the end of the testing." She would be calling him every thirty minutes, she said. And she very much hoped the result would be cholera. Not Ebola.

Augustine Goba now began his test of the blood; he would test only the blood of Mamie Lebbie, the woman in the clinic. He suited up and entered the Hot Lab. Meanwhile Moinya Coomber, Nadia's technician, began preparing a fresh set of all eight blood samples, for the second test in Nadia's machine.

While the work went on in the lab, Humarr Khan went outdoors to the alley next to the Hot Lab, sat down on a step, and lit a cigarette. He was in turmoil. The minister of health was breathing down his neck. The Ebola outbreak was thought to be almost over. But if *one* person with Ebola was found in Sierra Leone, it meant an outbreak was happening. How many people were out there in the villages right now, dying of Ebola? He lit a cigarette, smoked it, and lit another one.

8:40 p.m.

Nadia Wauquier was sitting in the Library, opposite the entry door to the Hot Lab, when she saw the door open a crack. Augustine Goba put his face up to the crack—he was wearing full PPE. "I think I have a positive," Goba called through his breathing mask, "but I can't tell."

Nadia went to the entry and suited up as fast as she could. Two minutes later she was dressed in full PPE and huddled with Goba over a piece of equipment called a gel box. Inside the box they could see a pattern of glowing green bands against a dark background. The pattern was the fingerprint of *something* in Mamie Lebbie's blood. Wauquier stared at the pattern. She couldn't tell if this was Ebola or not.

"Let's contact Kristian Andersen," Goba said to her. Andersen was a member of Pardis Sabeti's War Room Team, and he had brought the equipment to Kenema. There was no time to lose, no time to make an exit from the Hot Lab. Wauquier unzipped her suit, took her cellphone out of her pocket, and took a picture of the image. She opened up the email on her phone. "Hello," she typed. "We are in the Lab and have a question." She attached the photo and hit *send*, and waited. There was no reply. It was Sunday afternoon of Memorial Day weekend in the United States.

WAR ROOM

Kristian Andersen and his wife had spent the day hiking in the White Mountains of New Hampshire. They returned home shortly before five o'clock. He puttered around their apartment for a while, and at 5:30 he checked his email and saw the message from Nadia Wauquier. He looked at the photograph of green bands, which she'd taken forty minutes earlier in the Hot Lab. He thought it *could* be Ebola, but he wasn't sure exactly what he was looking at. This could be nothing. Or it could be a huge Ebola fire that was burning invisibly in Sierra Leone. Alarmed, Andersen sent Wauquier a reply asking for more information.

Meanwhile, Nadia had hurried out of the Hot Lab and gone into her container lab. There, her PCR machine was coming up with a new set of results on Mamie Lebbie's blood and the seven other blood samples taken from patients in the wards of the Kenema hospital. Once again, there were three positives, including the woman in the rural clinic.

Nadia felt a wave of dismay and embarrassment when she saw the second set of results with three Ebola positives. This was not a false result, it really was Ebola. *And it was in the hospital*. There were two patients in the Kenema hospital with Ebola disease. Kenema Government Hospital was already going hot.

She felt awful. She was a scientist. She was supposed to be a rational person. Yet she had rejected these Ebola positives the first time she had seen them. Her machine had been accurate. It had been screaming Ebola at her, but the mind doesn't see what it doesn't want to see.

Nadia walked outside her container and found Humarr Khan, still sitting on the step in the alley, in darkness. He stood up when he saw her. "So?"

"Your day is about to get worse."

"What—we have to redo the test again?"

"No." The woman in the upcountry clinic, Mamie Lebbie, really did have Ebola. And two more people had it, and they were inside the hospital.

"Oh my God." Khan sat down abruptly on the step.

At that moment, Pardis Sabeti was sitting at the dining table in her apartment in Boston's Back Bay, doing some work. Her pet rat was asleep in her lap. Her phone rang. It was Kristian Andersen, saying that he thought Ebola had gotten to Sierra Leone.

Thoughts flashed through her mind. There's never just one case of Ebola. It meant there was an outbreak happening. Invisible. Of unknown extent. It could be huge. Humarr Khan was in trouble. The Kenema hospital was in trouble. Sierra Leone was in trouble.

She called a meeting of the Ebola War Room team, to start in half an hour. Then she placed her rat on the floor, went to the refrigerator, and took out a carrot and piece of celery, and chopped them up, and placed little piles of the chopped vegetables at spots around the apartment, so the rat could have a treasure hunt while she was out. Then she put on a pair of roller blades and a helmet, and skated down Commonwealth Avenue and across the Massachusetts Avenue Bridge to Kendall Square, and carried her skates up to the Ebola War Room, on the sixth floor of the Broad Institute.

The War Room was a large room with whiteboard walls, which were forever covered with cryptic multicolored scribbles made by scientists writing with marker pens, a table made of blond wood, and a wall of windows that looked out on East Cambridge. About ten people had gathered around the table, scientists and post-docs from MIT, Harvard, and the Broad Institute. "Right now, Ebola is only a single blip on the radar screen of Sierra Leone," Sabeti said to the group. "It's only one confirmed case, but it means this is an outbreak. I dread what Humarr Khan and his people are facing."

Sabeti drew up a plan of action. The machines at the Broad Institute could read the genetic code of the Ebola that was hitting Sierra Leone and Humarr Khan's hospital. She knew that the virus's code was changing as the virus moved through human bodies. Mutations would be accumulating in the code of this Ebola as the virus amplified itself in human bodies. Just how different would the virus be by now? Was this West African Ebola starting to mutate into something different from classic Zaire Ebola? Something more contagious? Something harder to treat?

Sabeti and her team made plans to begin reading the genome of the virus as soon as possible. All the experimental drugs, experimental vaccines, and diagnostic tests for Ebola depended critically on the virus's genetic code. Could Ebola be evolving away from the tests used to identify it? Could the West African Ebola be turning into some kind of Ebola that not even the untested experimental drugs could possibly stop? Where had the virus come from? Had the outbreak started with just one person, the little boy? Or had the outbreak begun in different people at different times and in different places? Was Ebola attacking the human population from different starting points? The answers to these questions concerned every person on earth.

The War Room Team conceived a plan to obtain samples of blood from the people known to be infected with Ebola. They would read the genomes of whatever Ebola they could find in the patients' blood.

The virus was not a single thing but a swarm of particles. The swarm was moving through people and growing in size, and it could be mutating, changing its character as it grew. By looking at a few genomes of Ebola, the scientists hoped to make an image of the whole virus, which could be conceived of as a life form visible in four dimensions, as vast amounts of code flowing through space and time. To see the genome, they needed blood.

KENEMA
Monday, May 26

Nine hours later, Humarr Khan was standing in front of a window in the Library. The room was packed with staff of the Lassa research program. The air in the room was thick with warmth and anxiety. "Guys," Khan said, "Ebola is with us." He explained that two Ebola patients were in the hospital already. They were Satta K. and Victoria Yilliah; both women were now lying in beds in the Lassa ward, desperately sick. The Lassa virus surveillance team, he said, should prepare a group of vehicles and set off for Koindu immediately, the small town in the Makona Triangle where Mamie Lebbie, the first confirmed Ebola case in Sierra Leone, was lying in a bed in the local clinic. Get her and bring her back to the Lassa ward to get her isolated, and search the area for more Ebola cases, he instructed the surveillance team.

As Khan was issuing these instructions, a woman in the room started crying. Her name was Veronica Jattu Koroma, and she was the assistant supervisor of the Lassa ward—Auntie's deputy.

Khan spoke to her sharply in Krio. "Wetin make you de cry, Veronica? Why are you crying?"

"I am crying," she said to the room, "because I myself have gone through the arms of Lassa fever, and I know that Ebola is worse." Like the other nurses in the Lassa ward, she was a survivor of Lassa hemorrhagic fever. Her illness had been dreadful, and she had fallen into a coma and nearly died. As she had started to recover, she had gone bald. Lassa virus had killed the roots of her hair. She also fell into a serious depression. Years later, her depression had lifted and much of her hair had grown back, but she wore a wig to cover up the damage. "Ebola is more virulent," she said to her colleagues.

Khan wasn't sympathetic. He reminded her that she, like everybody else in the room, was a medical officer with a job to do. The Kenema Lassa program had the nation's only medical unit that could handle Ebola patients. Everyone in the room was an employee of the national government. "The Ministry of Health is looking at us," he said.

After the meeting, Veronica Koroma walked up the hill to the Lassa ward to start the day's work. In a few minutes, she would be coming face-to-face with Ebola patients for the first time in her life. She had never seen the disease, and she was very afraid. At the entrance to the ward she met her boss, Mbalu Fonnie. "Oh Auntie, I'm so afraid and so scared," she said, and started crying again.

Fonnie looked grave. "Veronica, why cry? Why pour tears? We have to put on our PPE and get to work on our patients." The two women went inside the cargo container next to the Lassa ward. It was the staging area for the ward. There, they donned whole-body Tyvek biohazard suits with hoods, HEPA breathing masks, eye shields, double gloves, rubber boots, and rubber aprons. Then the two women crossed the open space and went through the entry door into the hot zone, the narrow corridor lined with cubicles. From that moment on, what had been the national Lassa ward would be the national Ebola ward.

After the staff meeting, Humarr Khan dropped by the cargo-container office of Nadia Wauquier, the French scientist, to talk about the test results. While they were talking, he asked her if she could spare a cigarette. Khan then closed the door—he didn't want anybody to see him smoking—and continued discussing the blood tests. Suddenly Khan clapped his hand to his forehead. "Oh, shit!"

"What?" she asked

"Shit! The stool cups!"

The motorbike courier, after dropping off the tube of blood from the clinic, had delivered several cups of feces to a government lab in Freetown, to be tested for cholera. But the patient had Ebola. "Those cups are hot!" Khan exclaimed.

It was the world's hottest shit. They burst into morbid laughter, they couldn't help themselves, but it was a dangerous situation. Khan would have to call the Freetown lab and warn the workers not to open those cups, and to sterilize them.

But wait—how could a cup full of Ebola shit be sterilized? Nadia and Khan began discussing the problem. Should the cups be burned in an incinerator? The feces would spill into the fire, producing smoke. Could someone catch the virus by breathing shit smoke? They realized that nobody had any idea whether Ebola shit smoke was a biohazard. Khan called the Freetown lab and told them to burn the cups and to absolutely stay away from the smoke.

10:30 a.m., May 26

After smoking a cigarette with Nadia, Khan returned to the Lassa ward to meet with Auntie and discuss the clinical care of the two Ebola patients, Ms. Victoria Yilliah (who'd had the hemorrhagic miscarriage and the D&C), and Ms. Satta K. Satta K. had three children, two teenage boys and a young girl. They were sitting on a bench in Khan's outpatient waiting area, under the palm-frond shade, next to the Lassa ward, waiting for news of their mother. A thunderstorm came in, and Satta's children ended up in Khan's cargo-container office, sheltering from the rain and talking with him. They refused to accept the idea that Ebola was real or that their mother had the disease. Khan couldn't change their minds about this, and eventually he brought Satta's children to Nadia Wauquier's cargo-container lab to show them how the blood test worked. Nadia displayed their mother's results on a computer screen and explained how the test was done.

Satta's children listened attentively. They were bright kids. They told Khan that they wanted their mother's blood sent to a second laboratory in order to cross-check the results. Nadia later wrote in her journal:

> They seemed to be quite educated. I told them that . . . the only option would be to send it [their mother's blood sample] abroad, whether Europe or the US. I assured them they would get the same result. Then they started asking more moving questions

such as "what are the chances of survival?" "is there a cure or a vaccine?" I looked at Khan to see if he would help me but he shook his head and told me to go on and answer. Telling children that their mother has an 80% risk of dying is not an easy thing. They took it fairly well. . . . They thanked me sincerely for talking with them and moved out under the rain. I saw them again a few minutes later, outside. The young girl was crying.

Their mother died four days later, in the Ebola ward.

While Nadia was explaining Ebola to Satta K.'s children, the surveillance team was traveling into the Makona Triangle in a group of three 4x4 vehicles that included a bush ambulance. The vehicles were loaded with biohazard protection equipment. After hours of driving over mud-slogged roads, the team arrived at Koindu, the town where Mamie Lebbie, the laboratory-confirmed Ebola patient, was lying in a bed in the town's clinic.

It was Michael Gbakie's job to get the team dressed properly and to make sure they made no mistakes when they were inside the clinic and exposed to an Ebola patient.

First the team met with a group of Kissi chiefs—who have a lot of political power—and described Ebola to them, and explained what the team was doing. Michael speaks four languages—English, Krio, Mende, and Kono—but he doesn't speak Kissi. Nobody on the team spoke Kissi except for the ambulance driver, a Kissi man named Sahr Nyokor, who helped translate. Afterward, the team parked their vehicles near the clinic, a small, one-story building with yellow stucco walls.

The clinic was a hot zone. Inside there was one person who definitely had Ebola, and there might be other people in the building with Ebola, as well. Every interior surface of the building could have Ebola particles on it—walls, floors, beds, medical tools, toilets. As is normal in African medical care units, there would be family members in the building taking care of loved ones. The family members might be very protective of the patients. Any of the family members could have Ebola, too.

The team members began staging their gear. This was their first deployment into an Ebola-contaminated area, and they were keyed up. Michael supervised as they put on whole-body Tyvek suits, HEPA breathing masks, eye shields, double gloves, and rubber boots. Michael instructed the drivers to stay close to the vehicles. The drivers included the Kissi man, Sahr Nyokor. Then Michael and the others entered the building, which was now, they knew, a nest of Zaire Ebola, the red queen of the filoviruses.

VIOLENCE

KOINDU
2 p.m., May 26

As Michael Gbakie and his partners entered the clinic they scoped the layout of the place. There were four small ward rooms in the clinic, and the rooms were full of patients. The team found the Ebola patient, Mamie Lebbie, lying in a bed—a woman in her thirties, deathly ill, being cared for by anxious relatives, including her husband. The team asked the family to stop giving her care and not to touch her. They began exploring the clinic more carefully, examining patients, and to their surprise they discovered eight more patients who were showing symptoms of Ebola disease. Strangely, all of the suspected Ebola patients were women.

The sight of government medical officers in moon suits walking around the clinic frightened the patients and their families. Speaking Krio through his HEPA mask, Michael explained to Ms. Lebbie and her family that she had a disease called Ebola. It was very dangerous and could spread to other people, he said. He wanted to bring Ms. Lebbie to the government hospital in Kenema, where she could get treatment and could be isolated so she wouldn't spread the disease to other people. He had no legal power to force her to get into the ambulance. As a patient she had freedom of choice.

Ms. Lebbie, however, was too sick to make a decision. It was up to her family to decide. Her husband was in favor of moving her to the Kenema hospital, but her relatives objected. "The relatives raised their eyebrows at that idea," Michael recalled later. "They made references to the MSF [Médecins Sans Frontières/Doctors Without Borders] treatment center in Guinea. They said in Guinea people had been taken to the treatment center and it was the end for them."

Michael and the team decided to stay in their hazmat suits and try to reason with the family. They stayed for almost four hours, discussing the matter with Ms. Lebbie's relatives, but the relatives remained firm: She was to stay in the clinic.

Remember our earlier example in suburban Massachusetts. If you were visiting your sick mother at Newton-Wellesley Hospital, and a team of federal officers wearing moon suits suddenly entered your mother's hospital room and announced that an extreme virus had gotten into your mother and they needed to take her away to a government facility, you might have some questions about this. You might, in fact, start screaming.

As the discussions over Mamie Lebbie dragged on, a crowd began gathering outside the clinic, and rumors and text messages started flying around the villages in the area. As the crowd grew larger, Michael and the team began hearing the commotion. They could see people moving around outside the clinic's windows. The crowd was speaking in Kissi. Mixed in among them were many young men.

The drivers stayed close to their vehicles. Sahr Nyokor, the Kissi driver, understood what people were saying, and it scared him. He went over to a window in the clinic and began waving to Michael Gbakie, who went to the window and opened it a crack.

The youths, Nyokor told him, were planning to attack the team. They were going to burn the vehicles, so the team couldn't escape, and then they were going to move in on the team and try to hurt or kill them.

Gbakie rushed back to his team and told them to make a crash exit from the building, right now, get out of your gear, don't do decontamination procedure. Still wearing their moon suits, they ran out the front door and found themselves facing a group of hostile young men holding rocks in their hands. They were close to the vehicles, and had cut off the team's means of escape.

The team ripped off their masks and tore off their suits, and kicked their rubber boots off their feet. The drivers moved away from the vehicles and got in among the team members. Gbakie and his co-leader, an epidemiologist named Lansana Kanneh, exchanged quick words. They decided that their best chance was to try to reach the town's police station. It was four hundred yards away. They would have to run in their socks, having rid themselves of their biohazard boots.

As the young men closed in, the team broke out. They sprinted toward the police station, keeping together, and it turned into a desperate, four-hundred-yard dash. The young men went after them, hurling rocks. The rocks were the size of apples and came in on skull-fracturing trajectories. The team members, looking over their shoulders and dodging the rocks, made it to the police station and flew in through the door, gasping, while the crowd pulled up just outside. None of the team members had been hit.

The police officers seemed strangely unaware of the riot. Michael and the team told the police they wanted to file a report on the violent assault they'd just experienced. They also said they were afraid their vehicles were about to be burned. The police officers gave them a paper form to fill out stating the details of the incident, location, time, and so forth.

Michael filled out the report. But then the team couldn't leave the police station, because the young men were still around. Night fell, hours passed. They looked out into the darkness and wondered if they were going to see bursts of flame, their vehicles being torched. They heard motorbike engines sputtering through town.

Hours later, the town seemed to have cooled off. They returned to their vehicles, which were undamaged. During the hours when the team had been holed up, however, all nine of the sick women with Ebola symptoms had disappeared from the clinic, including Mamie Lebbie. Their beds were now empty.

What had happened was this: People had phones, people had motorbikes, and news had spread fast. The families of the sick women had organized a rescue. It takes about twenty minutes to drive by motorbike from villages near the Makona River to Koindu. Under cover of darkness, people riding motorbikes had come out of the villages and converged on the clinic, and had gotten their loved ones out. They put the nine Ebola-infected women on the backs of the motorbikes and carried them to safety, to be hidden in the villages or taken across the river to Guinea. Mamie Lebbie—it was later learned—was carried on the back of a motorbike to a river crossing, and was taken into Guinea. (Weeks later, she surfaced alive, and eventually gave some interviews to local news media—she was an Ebola survivor.)

The team couldn't spend the night in Koindu, the town had gotten too dangerous, so they drove to a larger, safer town and spent the night there. The next day they returned to the area around Koindu and began driving through the nearby villages, asking questions, looking for the nine women who'd been spirited out of the clinic. They were also looking for more people with signs of Ebola. Local people were reluctant to talk with them. And the villagers were clearly hiding the suspected Ebola patients.

KENEMA
8 p.m., May 27

After the team was on its way back to Kenema, Michael Gbakie called his wife, Zaiinab. He told her he was safe and that he'd be home from the patrol—the expedition—in time for a late supper. The children would eat early because they had school the next morning. The vehicles entered the hospital compound and pulled up next to the Lassa program office. Michael went into his office and unzipped his travel bag and dropped off some papers. He was exhausted. He took an L.L.Bean sweatshirt off a hook on the wall and put it on, to ward off the nighttime chill. On the wall of his office, over his desk, there was a large map of eastern Sierra Leone, dotted with hundreds of villages. He often consulted the map when he was tracking cases of Lassa fever. Right now, the map dotted with villages did not make him feel easy.

Ebola was out there somewhere, but it was going to be very hard to find. He slung his zipper bag over his shoulder, put on his helmet, and went outdoors and started his motorbike.

He wove his way around mud puddles, feeling the damp, clean air blowing over his face. The nights of the rainy season were beautiful, cool, and thunderstormy. His route took him past small houses crowded together, made of concrete blocks, separated by lumpy dirt streets. Many of the houses were dark, except for an occasional fluorescent bulb throwing a greenish glare. Inside the houses parents were cleaning up after supper or going to sleep, having already put their children to bed.

Years earlier, a civil war, called the Blood Diamond War, had devastated Kenema and the surrounding country. Marauding soldiers with automatic weapons had shot many citizens, had cut off their hands and feet with machetes, and had forced them to work in the diamond fields under threat of execution. The reason was diamonds. Control of the diamond fields equaled control of the nation of Sierra Leone. The war had ended and things were better now, but he often reflected that life in Kenema was not easy. It could be difficult to explain to his international colleagues what it felt like to raise a family in Kenema. He began following a road that runs alongside the dirt airstrip of the abandoned Kenema Airport. No planes had landed on it in many years.

He turned onto a dirt track that led to a group of houses that had been built on the end of the abandoned airstrip. He parked next to his house, a modest, spotlessly clean stucco structure painted yellow, with glass windows and a new metal roof. He and Zaiinab were raising two sets of children. Their own children were teenagers and young adults. They were also raising Michael's brother's children, who were youngsters. The sound of Michael's motorbike attracted the kids, and they came hurrying out of the house. He was an affectionate father. As he stopped the bike, the children gathered around him, expecting hugs.

"How de patrol? How de patrol?" the kids asked.

"I tell God thanks it was not so bad," he said. "No'r touch me"— don't touch me—he said firmly to the children.

The kids backed away. They had no idea why Dad didn't want

them to touch him. He stepped off the motorbike and took off his shoes and socks. The children watched him curiously. He had been walking around contaminated areas for two days, and his shoes could be hot. He nudged his feet into a pair of plastic slippers that he'd left outdoors intentionally. There could be Ebola particles on his feet, too. The slippers would keep his feet from making contact with any surface that the children might touch. Walking in the plastic slippers, he warned the kids to stay away from him, and told them to go indoors: "Now go back na house."

He stashed his shoes and zipper bag in a place where no one was allowed to touch them. Then he went to the well outside his house and drew up a bucket of water, and carried the bucket to a small structure made of concrete blocks which stood near his house. It was a wash house. He went inside, carrying the water.

BEDTIME

May 27

Michael Gbakie's wash house contained a decon room, a decontamination chamber that served as a makeshift gray zone, a barrier between the virus and his family. The room contained a plastic wash tub, now full of water, along with a scrub brush, towels, and packets of alcohol swabs. Right now, nobody was allowed in the decon room except himself.

He kicked off the plastic slippers and removed his wristwatch and set it on a dry surface. He then measured out a quantity of bleach and poured it into the water. Then he stripped off all his clothing and dropped it into the tub of bleach solution. He scrubbed his clothing with the brush, churning the clothes thoroughly. He would let them soak in the bleach water for thirty minutes. This was a validated kill time for Ebola particles. Next he slowly poured the bucket of water over himself, soaping his entire body carefully, rinsing himself, and he dried himself with a clean towel. He wiped his wristwatch with an alcohol swab to sterilize it, and put it on his wrist.

Naked except for his wristwatch, and now presumably free of Ebola particles, Michael Gbakie wrapped a towel around himself, got his feet into the plastic sandals, and walked across the yard and went inside the house. In the bedroom, he put on clean clothes. Then he went into the parlor, and at last embraced his wife and the children. Zaiinab had made a dinner of cassava leaves and smoked fish over rice, with okra. Everybody had eaten except for him, so he filled a bowl and sat down. They all wanted to hear about everything that had happened on his mission into the Ebola area. He told them how the villagers had attacked the team and chased them, and how their vehicles had almost gotten torched, and how they'd discovered and lost nine Ebola patients, and how people didn't believe that Ebola was real.

The children were spellbound. Zaiinab got really frightened, especially when he told her that he and the team would be returning to the Ebola area tomorrow, to search for more Ebola cases. As he went to bed, Michael had a strong feeling that the situation had already gone beyond anybody's control.

NEW ORLEANS
Simultaneously

While Michael Gbakie was telling his family about his trip into the red area, Lina Moses was standing in a hallway on the second floor of the Tulane Health Sciences Center in downtown New Orleans, looking at $60,000 worth of biohazard gear and medical supplies. The stuff was sitting in many, many cardboard boxes that were piled along nearly the entire length of the hallway. She was returning to Kenema immediately. She had been home for two days.

The American advisor to the Kenema Lassa program, a Tulane professor of microbiology named Robert F. Garry, had already flown to Kenema with a load of emergency supplies. He had authorized a large cash draw for Lina Moses and asked her to buy up every item of biohazard gear she could find on short notice, and then she was to bring the equipment to Kenema as quickly as possible. Humarr Khan and Auntie Mbalu Fonnie were going to need it, fast.

Lina felt a mixture of fear and exhilaration as she looked at the vast amount of gear she had collected. Stacks and stacks of disposable non-pressurized whole-body hazmat suits. HEPA breathing masks. Goggles. Tall rubber boots. Nitrile surgical gloves, the kind that are resistant to tearing. Rolls and rolls of tape. You need lots of tape for biohazard work, to seal cracks and to keep a hazmat suit tight. Many, many blood draw kits, for drawing blood from patients, blood to be tested in the Hot Lab for Ebola. Infusion lines, infusion needles, and infusion bags, for administering saline solution to Ebola patients, to help keep them from getting dehydrated. Pump sprayers, for spraying bleach disinfectant. Biohazard body bags, made of white Tyvek—there were going to be Ebola deaths.

Ever since she had been in college, Lina Moses had wanted to go up against Ebola in an outbreak. This had been her dream for years. Now it was really going to happen. It was a battle of a kind, a public health battle, and the aim was to save lives. This was Zaire Ebola, and nobody who went near it could consider themselves safe. Of course she could catch it, anybody could catch it, but Moses felt a sense of confidence in her training and experience. She was a seasoned epidemiologist, she had worked with Lassa virus for years, and Lassa was a Level 4 pathogen like Ebola. She spent the afternoon pulling things out of boxes and repacking the stuff tightly into twenty-seven plastic foot-lockers she'd bought at Walmart.

She got home late. Aron had picked up the girls at school and day-care, and he'd fixed them some supper. That evening, Lina and Aron cuddled in bed with their five-year-old, Audrey, and read a book to her. Their older daughter went off to her room to read one of her dystopian sci-fi novels. There was absolutely no mention of Ebola that night in Lina Moses's house. No mention about the fact that she was about to rush back to Africa to deal with Ebola. No use of the word *Ebola*. No talk about viruses at all.

As she lay in bed next to her husband, Lina Moses felt very protective of him, and of her daughters. She really didn't want the girls to know about Ebola. They were too young to be told much about the virus, and Aron wasn't much interested in viruses. He had a highly visual imagination. He had painted her portrait, and he understood her eyes the way no one else did, their mix of greens and browns, the flashes of emotion that would make her eyes fill with tears embarrassingly at the wrong moment. She did not want her husband to see in his mind's eye an image of her body and face displaying the symptoms of Ebola disease. Aron would stay at home and look after the girls and do his painting while she went to the front line.

FREDERICK, MARYLAND
Same night

Lisa Hensley carried James upstairs to his bedroom and kissed him good-night, and he climbed up the ladder into his platform bed, bringing his laptop with him. Ebola was blowing up in West Africa, and a number of new cases had suddenly appeared in Liberia. She would be going back on a second deployment to the lab at the former chimp station in Monrovia. She had been paying close attention to James, wondering if he was worried about her going back to the Ebola area. This time, he didn't seem worried at all. More like annoyed. To help cheer James up, she promised him that as soon as she got back they'd take a vacation trip to the beach in South Carolina, where they'd swim in the ocean and have a really great time together.

The next morning, Michael Gbakie left home early on his motorbike, bringing his zipper bag with him. He would be gone for several days. Cellphone service was spotty in the Makona Triangle, so he'd told Zaiinab that she wouldn't hear from him until he was on his way home and near Kenema, and then he'd call her from the road.

As she watched her husband's motorbike go down the road and turn toward the airstrip, Zaiinab Gbakie prayed for him. She was fearful that she and the children wouldn't see him again. He was a medical officer working in the national service, and his duty was taking him into a dangerous place where he could easily lose his life.

AMBUSH

During the next three days, Michael and the surveillance team prowled eastern Sierra Leone in several Land Cruisers, bumping along bad roads, stopping in villages, asking questions. Quickly they found twelve women who showed signs of Ebola; some of them had been patients in the clinic and had been spirited away on motorbikes while the team was taking shelter in the police station. All of the women had been at Menindor's funeral. The team transported all twelve women back to the Kenema hospital. Their blood was tested in the Hot Lab and came up positive for Ebola, and they were placed in beds in the ward, and were tended by Auntie and her team.

Michael Gbakie often traveled with his fellow surveillance officer Lansana Kanneh. At times they went with the Kissi-speaking ambulance driver Sahr Nyokor. A Kissi speaker was worth his weight in gold out in the villages. But driver Nyokor was also a bit of a problem. He came from Daru, a large town on the edge of the Triangle about an hour's drive from Kenema, and he had many friends and relatives in Daru. Quickly the team discovered several Ebola cases in Daru—the virus had already moved closer to Kenema. On May 29, Nyokor was driving an ambulance to pick up a suspected Ebola patient in Daru.

The drivers weren't required to wear PPE. They were supposed to remain inside the ambulance. The Ebola patient would be handled by team members who were suited up in full PPE. During the Daru pickup, however, Sahr Nyokor got out of his ambulance and went inside somebody's house for a visit with friends or relatives. He evidently didn't want to frighten his friends, so he didn't wear PPE. Later it came out that someone in that house had Ebola.

The day after Nyokor went inside the house in Daru, Lina Moses arrived at the international airport outside Freetown, along with her twenty-seven trunks full of biohazard gear and medical supplies. She got the stuff through customs and out to a Land Cruiser that was waiting for her in a gravel area that is always crowded with men offering to help travelers with their luggage. The men gathered around Moses, but she told them she could do things by herself, and she packed the Land Cruiser with the trunks. When the vehicle was full of trunks, she climbed up on top of the vehicle and started tying more trunks to the roof. A group of baggage handlers stood around watching. She threw a piece of rope around a trunk and cinched down the lashing.

"Now look at that woman's arms and shoulders," one of the men remarked in Krio.

"Oh, now look how strong she is!" another man exclaimed. They didn't realize she knew Krio. "Her husband is a lucky man," another man said. "He has such a strong, hard-working wife."

Moses smiled at them. She felt buoyant.

A few hours later, Moses was in the Library of the Lassa Laboratory, meeting with Robert Garry, the Tulane microbiologist who was the principal American advisor to the Lassa program. Garry had flown in from New Orleans a few days earlier, bringing with him a large amount of biosafety gear. They planned strategy. Moses would immediately set up a crisis operations center in the Library. She would serve as the nexus of epidemiology and operational support for Auntie in the Lassa ward, for Khan, and for the surveillance team. She would keep the Lassa ward supplied with biohazard suits and medical supplies, and she would coordinate communication and operations among the disparate parts of the Lassa program. The Lassa program was the main line of defense for Sierra Leone.

Robert Garry would pursue the virus in the laboratory. He was working closely with Pardis Sabeti, the genomic scientist at Harvard, and with Humarr Khan. They planned to assemble a collection of small samples of blood taken from Ebola patients and from people suspected of having Ebola. The blood samples would be shipped by air to Sabeti at Harvard, and Sabeti would lead the genome sequencing of the virus at the Broad Institute—to see how the virus was changing as it went from person to person.

Robert Garry's task in Kenema would be to collect and preserve the blood samples and get them shipped to Harvard; he would be working inside the Hot Lab. Humarr Khan and top officials at the Sierra Leone Ministry of Health were anxious to have the genome of Ebola sequenced, and so Khan and Sabeti, working with the ministry officials, would use a method of collecting blood that didn't interfere with patient care: The researchers would scavenge samples of blood serum from tubes left over from clinical care. This material was biohazardous medical waste. "We did everything we could to make no footprint in the way we took samples," Sabeti later said.

In addition to working in the Hot Lab gathering blood, Robert Garry would be traveling with Humarr Khan to community health clinics in the Makona Triangle. Their goal was to educate local health workers about Ebola. Khan and Garry also wanted to view the situation on the ground, up close. Moses and Garry worked out their plans in the Library, and Moses went to work delivering hazmat suits and supplies to Auntie at the Ebola ward. By then, the ward had fifteen Ebola patients in it. The ward had only twelve beds, but Auntie had brought in three extra cots. More Ebola patients were coming in all the time. Auntie and her nurses were wearing full PPE and working hard shifts, trying to keep up, while the patients were vomiting and having diarrhea, and dying. Moses found Auntie in the foyer of the Ebola ward. The two women embraced. It was the middle of the afternoon on the thirtieth of May.

Same time

Michael Gbakie and Lansana Kanneh, riding in an ambulance driven
by Sahr Nyokor, pulled into a village in the Kissi Teng Chiefdom
called Kolusu. Kolusu sits on a steep hillside, embedded in a deep patch
of forest. They had received information that one of the suspected
Ebola cases who had been taken out of the clinic on a motorbike had
died in the village. Accompanying them was an epidemiologist with
Metabiota, the American biotech firm, with a vehicle and a driver.
They parked in the forest near the village. The drivers stayed with the
vehicles, ready for a fast getaway in case there was trouble. Michael
and his partner advanced cautiously into the village. After some conversation, the villagers led them to a house where the body of a woman
was lying on a bed. Michael and his partner remembered her face. She
was indeed one of the nine women who'd been carried away from the
clinic. She had certainly died of Ebola.

They told the villagers that the body was dangerous, and should be
buried immediately with special safety precautions.

The villagers were hostile to this idea.

"They were in serious, serious denial. We talked to them at length,
and there were a lot of confrontations as it went on," Gbakie later said.
"They didn't believe at all in Ebola. It was not an easy conversation."

Gbakie and his partner stayed in the village for three hours, trying to persuade the villagers to bury the corpse. Night came, but the
epidemiologists made it clear that they weren't going to leave until the
body was buried. Finally two teenage boys stepped forward and said
they'd do the burial. Worried about the boys' safety, Michael and his
partner got them dressed in moon suits, and they dressed themselves
as well. Then they sprayed the body with bleach and got it inside a
biohazard bag. They placed the bagged corpse on top of an old wooden
door and carried the door with the bagged body on it to a place in the
ring forest where the villagers buried their dead, and they dug a grave,
sweating in their gear. By the time they had finished the grave, it was
nine o'clock at night. It was completely dark in the forest, and they
had no lights. They could barely see what they were doing. Just as

they were getting ready to lower the body into the hole, all hell broke loose.

The burial had been a setup. As they had worked, young men of the village had quietly snuck up and hidden themselves in the under-brush around the grave. At a signal, they began hurling rocks at the epidemiologists. Rocks the size of baseballs whipped past Gbakie and Kanneh. The two men ducked and yelled, and ran blindly into the for-est and up the steep hillside, trying to find their vehicles, while the young men chased them, throwing rocks. The attackers knew the for-est perfectly, and the white moon suits made good targets. Michael and his partner had no idea where they were going and they couldn't see anything in the forest. But then the attack stopped. They had shaken off their pursuers.

Gbakie heard a vehicle's engine start—and then it roared off down the dirt track that led to the village. It was the Metabiota team, getting away in a hurry. "They left unceremoniously," Michael later said. He and his partner could hear Sahr Nyokor's voice, and they ran toward it. He was shouting to them so that they could find him, and he had started his ambulance and turned it around, poised to make a getaway. They reached the ambulance just in time to be ambushed a second time. Rocks again flew out of nowhere, crashing into the doors. Nyokor gunned the engine, and the ambulance lurched forward as a rock punched a hole in the windshield, and more rocks smashed the side mirrors. Moments later they were out of the ring forest and moving fast.

They had almost been killed. They drove back to Kenema without stopping. Michael Gbakie got home at three o'clock in the morning, deeply shaken. He went into the decon room, sterilized himself, got dressed in clean clothes, and went into the parlor. Zaiinab and one of his older sons had stayed up waiting for him, and they were dreadfully worried. Zaiinab had kept dinner warm for him, but he was too upset to eat anything, though he drank some water and tried to get a few hours of sleep. He and the team, first thing the next morning, would be driving back into the Ebola zone to continue their search for people sick with Ebola.

He could now see the problem only too clearly: The local people didn't believe that Ebola was real. The virus was out there, it was spreading, and the local people would become violent if the team tried to find it. It was very clear to him—after nearly being killed in a village—that his nation was heading for a disaster. All he could do, personally, was just keep working and try to keep his family safe.

Robert Garry, the Tulane scientist, was working inside the Hot Lab, collecting blood samples from used tubes of patient blood. The Hot Lab was a small place, with a limited amount of equipment. Augustine Goba and his technicians were running a PCR machine, testing blood for the virus, so that patients who had Ebola could be distinguished from those who didn't. Nadia Wauquier, the French scientist with Metabiota, was running tests on her PCR machine in parallel with Goba and his machine: This was to ensure that each patient got two tests for Ebola. This practice would cut down on false results, which could be fatal for a patient. If somebody tested positive but actually didn't have Ebola, the person would be placed in the Ebola ward, where they would certainly catch Ebola. And if somebody tested negative but they actually had Ebola, the person would be sent home or placed in the general wards, where the person would spread the virus. Thus every blood test done by Augustine Goba and Nadia Wauquier was a matter of life or death. Both Goba and Wauquier, and their technicians, were under intense emotional strain, and they were handling raw infected blood. The slightest mistake with the blood and their lives could be over. Nobody working in the Hot Lab was sleeping much.

Robert Garry needed to use the equipment, and he worked at night, when the lab was less busy. Yet as time went by it became increasingly difficult for him to work ethically in the Hot Lab doing research, when the first priority had to be testing blood in order to save lives. In a few days, however, Garry collected scientific blood samples from forty-nine suspected Ebola patients.

The result was a large number of microtubes of human blood serum. Blood serum is a clear, golden liquid. It contains everything in blood except for the red cells, which have been removed. Each micro-

tube was the size of the sharpened end of a pencil and contained a
droplet of human blood serum no bigger than a lemon seed. The drops
each contained anywhere from a few hundred million to a billion particles of Ebola virus. The genetic code of the new Ebola was present in
the droplets, and was unread and unknown. The droplets were mixed
with a larger quantity of a sterilizing chemical that kills Ebola, and
were then frozen. Augustine Goba packed the tiny tubes of sterilized
blood serum on dry ice inside a box, and sent the box by international
courier to Harvard.

CAMBRIDGE, MASSACHUSETTS
June 4

Four days later, the box arrived at Sabeti's lab in Harvard's Northwest
Building, where a research scientist named Stephen Gire put on bioprotective gear and carried the box into a tiny biocontainment lab to
open it. The samples were supposed to be safe, but Gire was taking no
chances.

Gire is tall and quiet, and there is an air of seriousness and preci-
sion about him. After he'd gone into the biocontainment lab with the
unopened box of blood samples from Africa, he realized that he had
forgotten to bring along a knife. He opened the zipper of his suit, fished
his car keys out of a pocket, and slit open the box. The ice had melted,
but the tubes were still cold, and they were visibly safe: The color in
the tubes confirmed that the blood serum had been sterilized.

Gire's first job was to extract from the blood serum the virus's genetic material, and test all the samples for the presence of Ebola virus.
Of the forty-nine people whose blood samples were in the tubes, fourteen had been infected with Ebola. He could tell just by looking: In
those samples, the virus had damaged the blood, and the serum had a
murky look, clouded with dead red blood cells. Gire worked late, spin-
ning all the tubes in a centrifuge to clarify the liquid in them. He added
alcohol to the samples, and other chemicals. The Ebola particles in the
liquid fell apart, their protein cores breaking up, and the RNA in their
cores unspooled and came out, drifting like invisible hair in the liquid.
Using a pipette, which is a push-button instrument that is used for

transferring extremely small quantities of liquid from one place to another, Gire moved drops from tube to tube. The strands of RNA in the
liquid were delicate and brittle, like glass. As the drops were moved
around, the strands of RNA shattered into short threads.

When Gire was finished, he had fourteen small, clear droplets of
water solution, each in its own tube. Fourteen raindrops taken from
fourteen individuals who had had Ebola, all of whom had lived in the
Makona Triangle. In each droplet were vast numbers of broken strands
of RNA—shattered fragments of genetic code of the Ebola that had
once drifted in the blood of the fourteen people. There were many different genomes in the tubes, for the virus had mutated as it multiplied.

The next morning, Gire took a car to the MIT campus, carrying a
small box containing the fourteen samples of droplets with the Ebola
RNA in them. He parked and carried the box to the Broad Institute.
There, Gire and a colleague named Sarah Winnicki, working alongside two other research teams, prepared the RNA to be decoded. The
work took days, and was conducted in a glass-walled group of clean
rooms inside the Broad Institute. Gire and Winnicki hardly slept as
they worked on the drops, getting the liquid ready for processing in
a genome sequencing machine, which would read the code of all the
Ebola that had been collected from the Makona Triangle. They worked
with fourteen droplets separately at first. Then they combined the droplets, merging the Ebola code from the fourteen people into a mixture.

Pardis Sabeti stayed in touch with Humarr Khan, giving him
progress reports. He wanted to know how the decoding was going and
how soon it would be finished. Information about the virus's code
could perhaps tell him just exactly what kind of Ebola he was dealing
with in Sierra Leone or how it might be changing as it entered the
human species.

Sabeti told him that they didn't have a readout yet, but as soon as
she got any code from the test she was going to publish the results on
the Internet, so that scientists everywhere could get a glimpse of the
Ebola swarm as it changed in time. If anything significant showed up
in the code, she would let him know right away.

FLIP-FLOPS

EBOLA WARD, KENEMA
Early June

While the scientists at the Broad Institute were working with the droplets inside the clean rooms, the situation inside the Ebola ward was deteriorating. The beds were full, and as some patients died, more arrived. Auntie stood at the entrance of the ward, issuing instructions to her nurses in a whispery, British-accented voice, sending and receiving messages through staff, and sometimes putting on hazmat gear and going into the red zone to help the nurses and manage things.

The Tyvek material of the suit didn't breathe. The nurses got boiling hot and soaked with sweat inside their suits. In the tropical climate, you couldn't wear a whole-body PPE suit for more than about an hour before you were in danger of having heatstroke, which could be fatal. Auntie sent her nurses into the red zone in pairs, using a buddy system. A pair of nurses was a "hot team." While the hot team worked inside the ward, a nurse sat outside the red zone and watched a clock. At the end of an hour, the clock-watcher would order the hot team to make an exit, and another hot team would be sent into the red zone to work. This was like sending scuba divers into a dangerous operation and timing their dives.

The nurses were trying to be meticulous in their safety precautions, but they were getting scared. They didn't have any immunity to Ebola. The patients did not seem to be hemorrhaging much, but they were having explosive diarrhea and projectile vomiting. The ward was a mess. The suffering of the patients, and the way patients who seemed stable would suddenly crash and die in minutes, filled the nurses with horror and fear. Meanwhile the nurses' families were getting scared. The nurses were going home after working inside the Ebola ward, coming into contact with their children, their spouses, their parents. Many family members urged the Ebola nurses to stop working, and some of the nurses began skipping work, not showing up for their shifts. This tormented Auntie.

Downhill about fifty yards from the Lassa ward, in the Library room in the Lassa Laboratory building, Lina Moses was running the crisis operations center. The Library was directly across the hall from the entrance portal to the Hot Lab, and she could see the lab people going in and out, putting on gear and taking it off. The Library was stacked from floor to ceiling with boxes of biohazard suits and safety supplies. Moses sat at a table, taking and making calls on a cellphone, typing emails on her laptop, and meeting with a constant stream of lab technicians, staff workers, janitors, and surveillance officers. She often went on errands across the hospital grounds, and she usually ran, carrying medical supplies and safety gear uphill to the Ebola ward, and running back downhill carrying tubes of blood from the Ebola ward to the Laboratory building. She would hand the blood to somebody at the entrance of the Hot Lab.

Moses wore plastic flip-flops on her feet. She felt she needed to run from one place to another: There was always some emergency happening and somebody needing something fast. It was clear that Moses should not have been wearing flip-flops. She should have worn heavy rubber biohazard boots, especially when she went anywhere near the Ebola ward. A crowd of sick people and their relatives milled around the entrance of the ward, and some were infected with the virus. There were body fluids on the ground in front of the ward, vomit and feces. Moses refused to wear biohazard boots, because they wouldn't allow her to run. The flip-flops exposed the skin of her feet to the environment. She often ran up to see Auntie in the Ebola ward, her flip-flops slapping. Moses felt that she could tell where Ebola was and where it wasn't. Carrying a load of biohazard suits in her arms, she glanced at the ground and stepped carefully around messy spots, trying not to get anything on her bare feet. If a person looked unwell, she tried to stay six feet away.

Nadia Wauquier, who was working inside the Hot Lab and running blood tests with her PCR machine, became increasingly worried about Lina Moses. The two women were close friends. Nadia thought Lina had stopped paying enough attention to her personal safety in her efforts to help Auntie. Lina was always running to the Ebola ward, and her flip-flops really made Nadia nervous. She thought that if Lina got a small cut in the skin of her foot, or got a little bit of blood or vomit on her foot, she could end up getting infected. She decided not to say anything to Lina about her flip-flops, though. She had to trust that Lina wouldn't do anything stupid.

Humarr Khan was spending his time managing the crisis—hurrying around the grounds, rounding the general wards searching for patients with symptoms of Ebola, meeting with Auntie, meeting with the lab staff, meeting with the hospital's other doctors, meeting with families of patients, trying to encourage Auntie's Ebola nurses to keep going into the red zone. He worked closely with the district medical officer, an energetic doctor named Mohamed Vandi. Khan and Vandi were calling the Ministry of Health in Freetown, begging for more supplies, more assistance, more money.

The Ebola nurses were earning five dollars a day risking their lives in the Ebola ward. Khan and the district medical officer Vandi began asking the Ministry of Health for more money for the Ebola nurses. Government officials eventually agreed to provide each Ebola nurse with an extra $3.50 a day in hazard pay. But the money didn't show up. It was just a promise. Khan began to fear that the money was getting embezzled through corruption, or that the government bureaucracy couldn't be troubled to actually find any money for his nurses.

Khan and Vandi looked for international help, especially for more doctors who had any experience with Ebola patients. It turned out that there are very few doctors in the world who know anything about how to treat biohazardous patients who are hemorrhaging and rocket vomiting with a Level 4 virus. Khan got in touch with his friend Dan Bausch (who had talked him into taking the directorship of the Lassa program). Bausch was then working with the World Health Organization at a hospital in Conakry, Guinea, helping set up Ebola wards and bringing volunteer doctors into the fight as quickly as possible. Khan asked Bausch for help, and Bausch promised to send a WHO doctor to Kenema immediately. Bausch also promised to send additional WHO Ebola doctors in a few weeks, as soon as he could possibly get some. He added that he would also go to Kenema to help his friend Khan as soon as he could.

On June 8, Dan Bausch's first WHO doctor arrived at the Kenema hospital. A Land Cruiser stopped in front of the Ebola ward, and a rugged-looking thirty-something man with a shaved head and a sparse beard stepped out of the vehicle and asked for Humarr Khan. He was a British doctor named Tom Fletcher, a virus researcher at the Liverpool School of Tropical Medicine, in Liverpool, England. Fletcher, an expert in delivering clinical care in Ebola outbreaks, was volunteering for the WHO as a kind of advance special operative. He went into chaotic, Ebola-ridden hospitals ahead of the Ebola doctors, where his mission was to stabilize the hospital and make it safe for the Ebola doctors who would follow him. Fletcher was carrying a single box of medical supplies. "I was worried about Khan. I knew he was getting tired," Fletcher later said.

Khan arrived, and the two doctors talked briefly. They had never met. Fletcher sized up Khan quickly, and thought he seemed competent and dedicated. The two men then went into the Ebola ward to have a look, first suiting up in the cargo container staging room, where they inspected each other's hazmat gear. Fletcher noticed that Khan wasn't perfectly smooth at putting on his gear. Then they went through the door into the red zone.

There were fifteen Ebola patients in the cubicles along the narrow corridor. Fletcher could see that the nurses were under pressure. "They were a pretty frightened, tired group," he recalled later. Khan told Fletcher that some of the Ebola nurses had been skipping work. They were afraid of catching the virus, and their family members had been pressuring them to stay home so they wouldn't infect their families.

Patients were prostrate, and they were vomiting and having diarrhea. The nurses were giving them fluids to drink, but the patients vomited them up, which caused the patients to become severely dehydrated. As this happens, the level of potassium in the bloodstream decreases drastically. An imbalance of potassium in the blood can trigger a heart attack.

Khan was very concerned about keeping the Ebola patients hydrated. For years in his own practice he had prescribed coconut water to his patients. Coconut water was cheap, poor people could afford it, and it was rich with salts and minerals. But the patients were having trouble keeping liquids down, and kept throwing them up.

The alternative was to give an Ebola patient an intravenous infusion of saline solution, which could quickly bring the patient's fluid and potassium levels back to normal. There was plenty of saline on hand at the Ebola ward, with plenty of infusion kits. But to put a needle into an Ebola patient's arm seemed extremely dangerous—the worker could get pricked. The Ebola teams from the International Red Cross and Doctors Without Borders didn't ordinarily give IV saline infusions to Ebola patients—the procedure was thought to be risky, because a medical worker could get stabbed with a bloody needle. Khan and the nurses were following a standard international policy of not using needles in the red zone.

Tom Fletcher had a trick for setting up an IV safely. He showed the nurses a technique for safely placing an infusion needle in an Ebola patient's arm without endangering the nurse. The trick consisted of flipping a plastic cap over the needle so it wouldn't prick you. From that day on, and with Khan's encouragement, the Kenema Ebola nurses began giving IV saline infusions to all Ebola patients. "The nurses were phenomenal, really," Fletcher said. "They were trying to deliver high quality care, sticking IVs into everybody, doing their absolute best for the Ebola patients." It gave him a sense of confidence. "I was pretty hopeful. 'This is pretty good,' I thought."

That evening, Fletcher and Khan got dinner at a hotel in town, drank a beer, and worked on a strategy to stabilize the Ebola ward. After dinner, the two men returned to the hospital, put on PPE, went into the Ebola ward, and worked into the night.

Days went by. Khan suited up and rounded the red zone, and he and Fletcher worked together. Every evening, the two doctors ate dinner at a hotel, drank a beer, planned action, and returned at night to continue working. Fletcher's respect for Khan grew. They became friends. Meanwhile more and more Ebola patients kept arriving, including children. Auntie kept bringing in more cots, until the ward got so jammed with cots that it became difficult to move around. Patients were dying in the beds and in the cots, and the nurses were putting the bodies into biohazard body bags and removing them. Then there was the problem of food. The patients, especially children, needed to eat, if they could hold down food. Khan and Fletcher worked on providing a steady supply of food to the ward.

Khan and Fletcher remained optimistic that they could get things under control, but there wasn't any way to control the virus outside the hospital. Four days after Fletcher's arrival, there were twenty-five Ebola patients in the ward. Furthermore, Fletcher and Khan knew that there were additional undiagnosed Ebola patients hidden in the hospital's general wards. The symptoms weren't always obvious. Ebola was a disease with different faces, and in its early phases it could look like malaria or dysentery. As Khan went on his morning rounds, he kept finding people who had the symptoms of Ebola. He ordered blood tests, and some of the patients tested positive, and were sent into the Ebola ward.

The nurses in the general wards—who had no biohazard protection and zero training in it—couldn't tell who had Ebola and who didn't. To the general nursing staff, it seemed as if the virus could be anywhere at the hospital, or everywhere.

It was clear that the Ebola ward wasn't large enough to hold the growing number of patients. Khan and the district medical officer Vandi began work on constructing a plastic tent that had been donated to the hospital by Doctors Without Borders. Just as the structure was being finished, a rainy-season thunderstorm destroyed it. Khan and Vandi immediately set out to build a larger tent. In the meantime the Ebola ward got packed. Lina Moses was running around in flip-flops tending emergencies. She got too busy to take her malaria pills, and she broke with malaria. Shaky and feverish, Moses continued to manage the crisis center. Michael Gbakie and Lansana Kanneh, prowling around the Triangle in ambulances, continued to find and bring in fresh Ebola patients. Tom Fletcher and Humarr Khan began to fear that they weren't going to be able to stabilize Kenema Government Hospital. And then nurses who worked in the general wards began leaving their posts, becoming fearful that the virus was getting into the general wards. The general staff had started abandoning the hospital.

At this point Khan realized that he and his own staff in the Lassa program were getting cut off by the virus. The virus was driving away the hospital's main staff, leaving Khan and his people increasingly exposed. The virus was out there and growing, and it was hitting the hospital harder and harder, and starting to erode the medical system in Kenema the way a rising tide washes away a sand castle. Khan and his people were still at their posts, but it wasn't clear for how much longer they could hold against the virus. Meanwhile the rest of the world hardly knew that Ebola had broken out in West Africa, and the world hadn't noticed Kenema hospital at all. It was a small, forgotten hospital hidden in the diamond fields of Sierra Leone, where something very bad was happening. Khan decided he had to act, and he called a meeting of the entire Lassa program staff, with mandatory attendance.

SPEECH

Khan's meeting was held at the crisis operations center—in the Library, next to the Hot Lab. Virtually everybody in the Lassa program showed up—epidemiologists, nurses who weren't on duty just then, lab technicians, drivers, janitors. An air conditioner was running but the room was boiling hot. People stood pressed against one another, silent and apprehensive.

Khan spoke in a subdued voice, in the English of Sierra Leone. "Gentlemen, this is a tough encounter with Ebola," he said. "This is a very tough battle. Extraordinary things are happening, and we have extraordinary things to do." The nursing staff was abandoning the hospital, yet the general wards were still full of patients. Family members of the patients were taking over the care of their loved ones in the absence of nurses. The virus had gotten into the general wards. People were very, very scared.

"Since most of the health workers have run away from the hospital, we should be ready to work," he went on. "If you are working eight hours a day, be prepared to work many more hours. The ministers of government are very definitely looking over us now—the government is watching us closely." What the Lassa staff had to do was to stay in the fight against Ebola, and not desert their posts the way the general staff was doing. The rest of the hospital might collapse, but Khan's team had to hold its ground.

Somebody in the room began crying softly. Other voices joined in the crying.

Khan spoke over the crying. "This is our work to do. This fight is our fight now. We are working for our nation."

More people began weeping.

Khan raised his voice. "If you say you will not do the work, who will do it? We must do what we can as a national sacrifice."

When Khan used the term *national sacrifice*, the crying went all through the room. The staff of the Lassa program knew exactly what he meant. He was predicting that some of the people in the room were going to die. They were government employees. They worked for the Ministry of Health of Sierra Leone. If any of them died, it would be a sacrifice for the nation. Yet as the staff members of the Lassa program looked around the room, they could see how small their team was. They filled just one room, and they were the only people in their nation who were trained to deal with a hemorrhagic fever virus. They were the front line. As they looked at one another, they had no way to know who would make the sacrifice, who would die. Casualties were coming, and Khan was warning them so.

"No, don't cry," Khan said. "All we can do in this fight is take the precautions. Just be cautious." The room cleared out and the staff went back to work.

One member of the Lassa staff wasn't at the meeting. It was the Kissi ambulance driver Sahr Nyokor. He was at home spitting up blood.

TEARDROP

Working inside the glass-walled clean rooms, the scientists at the Broad Institute had combined all fourteen droplets of Ebola RNA from fourteen individuals who lived in the Makona Triangle. The result was a single, crystal-clear droplet of water solution. It was the size of a raindrop, and it contained about six trillion snippets of DNA. Each was a mirror image of a piece of RNA from the blood samples collected from the fourteen people. Most of the snippets of DNA in the droplet were human genetic code—bits of DNA from the fourteen people—but among the bits were about two hundred billion pieces of code from Ebola. There were also many billions of fragments of code from bacteria and other viruses—from anything that happened to be living in the blood. This droplet was referred to as a "library."

Each piece of DNA in the droplet had been tagged with a unique bar code—a short combination of eight letters of DNA code—identifying that particular fragment as having come from one of the fourteen patients. "You could consider each bar-coded fragment of DNA as a kind of book," Stephen Gire said. "The book is bound in covers and has an ISBN [International Standard Book Number] on it. It's a short book, so a reader can easily digest it. You can find the book by its ISBN number, and that's why the droplet is called a library. The books in the DNA library are bound so that the library can be put in a machine"—a genetic sequencer—"and the machine reads all the books." The "books" of DNA letters were all sitting in one immense, jumbled pile, and what was between their covers was unknown. Although the droplet was just a spot of water with DNA in it, it held as much information as the books in fifty thousand Libraries of Congress. This shows the ability of life to store huge amounts of information in a very small space. The library droplet contained fourteen images of Ebola virus, fourteen frames of a movie taken of the virus as it began to chain its way into the human species. The images were jumbled into tiny fragments and mixed together with vast amounts of other fragments inside the library droplet. The fourteen images still had to be found and pulled out of the droplet.

On Friday, June 13, Gire carried a single microtube containing the liquid droplet library to a logging station in the Genomics Platform of the Broad Institute. He left the tube sitting in a box, and logged in a request to have the droplet sequenced as soon as possible. The task was to read the genetic information in the fifty thousand Libraries of Congress in the droplet, and thereby start developing an image of the shifting code of the Ebola swarm as it flowed through human bodies in West Africa.

In Cambridge, at the logging station at the Broad Institute, a technician picked up the box that contained the small tube with the crystalline droplet inside it—the library of genetic code taken from the Ebola swarm in the Makona Triangle. The technician carried the droplet out of the building, down the street and around a corner, and arrived at a low, mud-colored building that had once been a storage facility for the peanuts and beer that are sold at Fenway Park. The building is now owned by the Broad Institute, and it contains the most powerful array of DNA sequencing machines in the world. Sixty of the machines sit in the center of the former peanut-and-beer warehouse, lined up in rows.

The machines, tended by crews of operators, run twenty-four hours a day, seven days a week, reading letters of DNA extracted from biological samples.

At the time the Ebola blood samples were decoded, each DNA sequencing machine was the size of a chest freezer and cost a million dollars. At the time of this writing, the Broad's DNA sequencing machines are the size of a desktop printer. They still cost a million dollars apiece, and they are still housed in the ex–Fenway Park peanut-and-beer storage warehouse. Somehow, the Broad scientists haven't gotten around to moving their most important equipment into the expensive crystalline buildings around the corner.

Recently, the machines have been reading the DNA of human genes involved in schizophrenia, autism, obsessive-compulsive disorder, major depression, and childhood allergies. The Broad's machines are also involved in a project to understand how each and every protein in every cell in the human body operates. The machines have been reading the DNA of cancer cells—part of a long-term effort to learn how to kill *any* cancer cell in *any* patient. The machines have been reading the code of the entire human microbiome—of every kind of bacteria that lives on or inside the human body. The human microbiome lives in the intestines, in the sinuses, in fingernail grime, dandruff, tongue fuzz, tooth plaque, earwax, elbow sweat, foreskin smegma, belly button lint, and toe cheese.

The Broad's machines have sequenced the genome of tuberculosis bacteria, of the malaria parasite, and of the mosquito that carries malaria. They've read the DNA of the coelacanth, of the rabbit, and of 4,400 skeletons of people who lived in various places around Europe during a fairly interesting period of the Bronze Age that occurred shortly after Stonehenge was completed.

Back to the raindrop of Ebola. In the room of the machines inside the peanut-and-beer building, a technician, using a pipette, sucked up about a tenth of the tiny library droplet—an amount like a fleck of moisture on a wet day—and placed it on a glass slide known as a flow cell. The fleck of liquid contained the full library of code from the blood of the fourteen Ebola patients from the Makona Triangle. The bit of water spread into channels on the flow cell, which sat in the mouth of a sequencing machine.

For the next twenty-four hours, the sequencer worked automatically, pulsing liquids across the flow cell, while lasers shone on it. On the surface of the flow cell, hundreds of millions of fragments of DNA had gathered into hundreds of millions of microscopic colored spots. The colors of the individual spots were changing as the process went on, and a camera took pictures of the changing field of spots and stored the data. Twenty-four hours later, the machine had finished reading Gire's library of bar-coded fragments of DNA. The data were sent to the Broad Institute's computer arrays, which assembled all the fragments into finished genetic code—it organized the vast pile of books in the library and placed the letters of all the books in their proper order on shelves.

Late afternoon, Sunday, June 15

Gire and Sabeti got word that the computers had finished their work. The result was twelve full genomes of Ebola virus—the Ebolas that had "lived" in the bodies of twelve of the fourteen people. (The computers had not been able to assemble the Ebola genomes of two of the people.) Sabeti and her team started the work of analyzing the code, to see how Ebola was changing. They printed out sequences of letters of Ebola code, and began staring at the letters, looking for patterns. They worked until dark that day.

As the sun went down on the East Coast of the United States, in West Africa it was night, and Lina Moses was working late in the crisis center. Abruptly she heard the sound of a diesel engine starting—it was one of the ambulances, she thought. The ambulance crews were forbidden to go out after dark because people were hostile to the ambulances and the roads were getting dangerous. But this ambulance was going out at night. Something bad was happening, but Moses had no idea what it might be.

The ambulance crew had gone out to pick up one of their own. It was Sahr Nyokor, who on two separate occasions had saved Michael Gbakie's and other team members' lives. It was Nyokor who had heard the crowd talking by the clinic and warned Michael they were getting ready to attack, and Nyokor who had driven the getaway ambulance at Kolusu amid an ambush. Now Nyokor was heaving blood. Somebody at his house had called for an ambulance.

His fellow drivers took him to the best ward at the hospital, called the Annexe ward. It was a private ward, where patients paid extra for their care, and was situated next to the Ebola ward. He was admitted to the Annexe late that night. Almost as soon as he got settled into bed, he started feeling better; he had stopped throwing up.

SWARM

It was the next day that Dr. Tom Fletcher, the advance operative from the WHO, understood that his mission was going to fail. He was not going to be able to stabilize the Kenema hospital and prepare it for Ebola doctors. Fletcher discovered twenty-eight new Ebola cases in the town of Daru, on the outer edge of the Makona Triangle and an hour's drive from Kenema. Twenty of the cases were in the Daru community clinic, which had gotten flooded with Ebola, and another eight people were found sick or dead in their houses in Daru. Fletcher had been optimistic, thinking he could help Khan get control of the situation, but now he saw that the virus was coming out of the Makona Triangle in a wave, inside people. People riding motorbikes, taking jitneys and taxis, arriving at the Kenema hospital, going to stay with relatives, going anywhere for help, heading for Freetown. The road to Freetown went from Daru through Kenema. Fletcher foresaw that a wave of Ebola was going to come out of Daru and sweep through Kenema in about a week's time. The coming Ebola wave was likely to overwhelm Humarr Khan and his people. The virus was no longer under anybody's control. It had gone beyond human control and had become a force of nature.

Fletcher had to leave for important work elsewhere; he had been sent to Humarr Khan on a short assignment and he couldn't stay. He called the WHO and asked for several Ebola doctors to be sent to Kenema immediately to give Khan backup. But there weren't any doctors available to help Khan. The doctors who knew anything about Ebola were busy fighting the virus elsewhere in West Africa. Fletcher did get a commitment from Dan Bausch, Khan's friend, to send two Ebola-veteran WHO doctors to Kenema in about two weeks. One of those doctors would be Dan Bausch himself. But there would be a gap of about two weeks during which Khan would be alone at the Kenema hospital. Fletcher feared that chaos could overwhelm Khan and his people during those two weeks. The wave of Ebola was coming.

Fletcher hesitated; he thought he should stay with Khan for an extra two weeks while Khan waited for backup doctors. Fletcher phoned his bosses at the Liverpool School of Tropical Medicine. "I'm having a really hard time walking away from this."

He was deeply worried about Humarr Khan. And when he departed, Lina Moses and Nadia Wauquier would be the only two foreigners left working at the hospital. He felt that their lives were in danger.

Nevertheless, on June 17, Tom Fletcher loaded his backpack into a Land Cruiser and said goodbye to Moses and Wauquier. "It's going to get a lot worse," he said to the women in a shaky voice. He embraced them, and they thought his eyes were wet.

"I was close to tears," Fletcher said later. "It was very difficult leaving. We couldn't send in just kids from the WHO to help Khan. He needed doctors who were experienced with Ebola." Fletcher didn't know if he would ever see Humarr Khan or the two women alive again.

The women watched Fletcher's vehicle go slowly up a hill along a dirt road, heading for the hospital gates. "Lina and I felt completely abandoned," Nadia Wauquier later recalled. "We didn't know when any others were coming to help."

After Fletcher left, Humarr Khan phoned Pardis Sabeti in Cambridge. "I feel all alone here," he said to her. "We need more resources.

We aren't getting the help we need. All the aid organizations are entrenched in Guinea. We need more foreign aid and more doctors working at Kenema."

Sabeti thought Khan sounded desperate, and it made her feel desperate. The War Room group was growing larger by the day, but she felt impotent, unable to help him. She also felt close to the nurses at the hospital. She had visited the hospital and had been deeply impressed by them. The ties to Kenema were strong. But all the DNA sequencing machines in the world weren't going to help Humarr Khan and those nurses. Feeling afraid for Khan, she tried to give him a sense that she and the War Room group were doing everything possible to get more doctors to Kenema. "Know that we are working to get you help, Humarr. We're calling all over the place." But Sabeti wasn't getting results with her phone calls. It was easy to get a promise of help from an organization or a government, but it was extremely difficult to get any actual help.

Sabeti's colleague Robert Garry, who had collected the blood samples in the Kenema Hot Lab for Sabeti, had flown to Washington to try to get some U.S. government help for the Kenema hospital and for Sierra Leone in general. He had just left Kenema. He knew what was happening in the Ebola ward, and he had visited Daru himself and had seen with his own eyes some of the twenty-eight people with Ebola who had just been found there. Like Tom Fletcher, Robert Garry could see the explosion of cases happening in real time, and he knew that an Ebola wave was heading for Kenema, and he tried to warn people in the U.S. government. "I went to a bunch of places in Washington," Garry said later. "I went to Health and Human Services, I went to USAID, I talked to people in the State Department and at the NIH." He was going around Washington just as the World Cup soccer matches were taking place. "I was giving a seminar on the Ebola situation, and maybe I'm a bad speaker, but I couldn't help but notice people checking the soccer scores on their phones. God, they could have taken it a little bit more seriously." In the end Garry couldn't get anybody in the U.S. government to arrange actual immediate help for Kenema, or, he thought, to consider the possibility that a crisis in a small hospital in Africa might actually be a crisis for every person in North America. "They finally took it seriously when Ebola got to a Dallas hospital," Garry said.

There was a widespread view among public health experts that Ebola "burned itself out" when it entered the human species. The virus was too hot, too lethal; it killed people too quickly to be able to establish itself as a permanent disease of humans. This was the widespread opinion, anyway. The simple fact is that Ebola virus just wasn't perceived as a serious threat.

The other reason Pardis Sabeti and Robert Garry couldn't get help to Humarr Khan was because of a shortage of medical people who had training and experience at dealing with an outbreak of a Biosafety Level 4 hemorrhagic fever virus. There just weren't enough knowledgeable doctors. Doctors Without Borders had taken the lead in crushing Ebola during past outbreaks. The Doctors knew how to set up an Ebola biocontainment ward and run it safely. They had biocontainment tents, they had spray pumps, laboratory equipment, generators, food, doctors, and robust supply lines, and they had a depth of operational experience with Ebola. And the Doctors were already stretched to their maximum. The medical world as a whole had no idea, no clue, how to stop an Ebola outbreak or how to safely handle patients infected with a huge, aggressive, Biosafety Level 4 virus.

"I'm worried about your stress," Sabeti said to Khan. "The most important thing is your safety, Humarr. Please take care of yourself."

"I feel I have to do all I can," Khan answered. He asked her about the genome sequencing of Ebola. Was the virus mutating? Was it becoming more lethal than Zaire Ebola? Khan, Lina Moses, Mbalu Fonnie, and the nurses had been struck by the features of the Ebola disease in the Lassa ward. There seemed to be less bleeding, less frank hemorrhage, but there was a huge amount of vomiting and diarrhea. Lina Moses wondered if this new Ebola was more transmissible than Zaire Ebola, because the patients produced huge amounts of infective fluids, which got splashed and smeared all over the nurses' protective gear. The patients did bleed from their gums, and their urine could turn bloody, but the nurses didn't see nosebleeds. Of course there were expulsions of melena—hemorrhage—from the intestinal tract. Was there something different about this West African Ebola? Was it really the same as classic Zaire Ebola?

And finally, was it mutating? There was a widespread view among Ebola experts that Ebola virus doesn't evolve in humans. Ebola, they asserted, was very unlikely to mutate significantly during an outbreak. Pardis Sabeti, looking at the Ebola code that had been collected from the twelve people in the Triangle, could see that the Ebola *was*, in fact, mutating. Its letters had changed, here and there, in the twelve Ebola genomes she was looking at. But there was no way to tell whether the mutations were just noise, only meaningless random stuttering of the Ebola genome as Ebola passed through people, or whether Ebola was evolving and getting to know people better.

Whatever the virus was doing, it was definitely a swarm. As of June 18, the growing swarm of Ebola particles in West Africa was still very small. Only about four hundred people, at most, were infected with the virus. Each infected body contained anywhere from a hundred trillion to several quadrillion particles of Ebola virus. In all, the swarm consisted of around forty quadrillion to a quintillion particles of Ebola.

A quintillion looks like this in digits: 1,000,000,000,000,000,000. It is a small number by the measure of a virus. A really small number. At this point, in the middle of June, the swarm was only the beginning of the entity it would soon become.

There are 18,959 letters in the Ebola genome, arranged in an exact spelling. As each particle replicated, there was a certain chance that an error in a letter would occur, and the spelling of the genome would change. Many of the changes in spelling didn't change the character of the virus itself. But there were misspellings that could change the virus by a lot, very suddenly. The growing Ebola swarm could be thought of as a huge, invisible biological pachinko machine with a quintillion balls bouncing in it; a cloud-like, expanding entity that was flooding into the human species along many chains of infection, all the while making vast numbers of random tests of human bodies to see how best to penetrate them and move through them and make itself immortal in them. Sabeti and her colleagues felt a definite fear that the Ebola cloud could get a mutation that would change it very suddenly, and would make it better adapted to the human body. They were trying to make an image of the thing. And it had come from a child who had touched something that lived near his house and got a few of the particles in his bloodstream.

Khan was dealing with the entity. He told Sabeti that it was hitting his hospital hard, and that somehow he and his colleagues hadn't seen it in Sierra Leone when it first arrived. "How could we have missed it? *How* could we have missed it?" Khan said to people over and over. He asked Sabeti if she could tell whether the virus was changing. She said that she and her colleagues were still analyzing the data. They didn't yet know how the genetic code of Ebola was changing, or if it was changing significantly. As soon as she knew anything, she would tell him, she said. She urged him to keep himself safe.

KENEMA
Ten hours later

Before dawn on the morning of June 18, Dr. Azziz Jalloh, who was still working in the general wards, did regular rounds in the Annexe ward. There, he came across the ambulance driver Sahr Nyokor. Mr. Nyokor had been admitted with a suspected bleeding stomach ulcer. Dr. Azziz found the driver "in a tormented state," as he later recalled. Nyokor was mentally confused and writhing with abdominal pain.

Dr. Azziz wondered if the man could have Ebola. But the symptoms didn't add up. The driver had only a slight fever, and he hadn't vomited once since his admission. No diarrhea, either. In fact he was constipated, and hadn't had a bowel movement in two days. Dr. Azziz examined the inside of Mr. Nyokor's mouth for telltale signs of Ebola. He was looking for inflammation inside the mouth or for hemorrhage coming from the gums. Mr. Nyokor's mouth appeared perfectly normal. In fact, Mr. Nyokor was in the false dawn of Ebola and was about to die.

Despite the fact he couldn't see any Ebola symptoms, Dr. Azziz had a gut feeling that the man had Ebola, and he ordered a blood test. As Dr. Azziz examined the ambulance driver, a nurse was standing by. In this book she will be called Lucy May. (Her true name has been withheld for the sake of her family's privacy.) Dr. Azziz gave some instructions to Lucy May about caring for Mr. Nyokor, and he left the ward at about six o'clock that morning.

Lucy May continued to care for Mr. Nyokor. She was thirty years old, a married woman with a delicate appearance, a devout Catholic who styled her hair in a modest bob. Lucy May had a beautiful singing voice, and she sang in the choir of St. Paul's Cathedral in Kenema. Unlike many of the nurses, she had continued working in the Annexe ward even after other nurses had gotten fearful of the virus and had stopped coming to work. It is pretty clear that she stayed at her post supported by the strength of her religious faith.

About an hour after Dr. Azziz left the ward, at about seven a.m., Nyokor got out of bed and walked to the toilet at the end of the ward. While he was in the toilet, he had a bout of diarrhea—his constipation had suddenly ended. While he was in the toilet he collapsed and fell, and hit his head on something. Nurse Lucy went into the toilet to assist him and discovered that he was bleeding from a wound in his scalp. She cleaned the blood from the laceration on his head and got him back into bed. An hour later, Nyokor went into sudden shock and died abruptly.

We can imagine, but don't know for sure, that Lucy May stayed with Nyokor as he died. The two were colleagues on the hospital staff, and she probably knew him personally, at least a little bit. It's easy to imagine that when she realized that her colleague was dying she prayed for him, asking God to have mercy on him in eternity, and maybe she held his hand as he passed. A few minutes after the ambulance driver died, the nursing shift ended. It was eight o'clock in the morning, the time of the shift change, when the night nurses went off duty and the day nurses came on duty. Lucy May left the Annexe ward and went home to get some rest. She needed her rest, because she was pregnant and her baby was due soon.

BLOOD DRAWS

8 a.m., June 18

After Nurse Lucy May went off duty at the Annexe ward, a nurse named Iye Princess Gborie came on shift in the ward and took Nurse Lucy's place. Nurse Princess was a tall, handsome woman, who would sometimes get a faintly skeptical expression on her face, and she was a Christian. She wore a small golden cross around her neck on a golden chain. The ambulance driver had just died, and it was her responsibility to do the last offices for the body. Like Nurse Lucy, Nurse Princess had remained at her post as Ebola appeared in the general wards. If Ebola were to appear in the general wards of any hospital in the United States it would be a national emergency. It was the same at the Kenema hospital, a national emergency, and the nurses knew this. It's why Princess Gborie had reported to work that morning, though by all accounts she was deeply afraid of the virus.

She closed the ambulance driver's eyes. She may have known him personally or at least recognized him. If there was any remaining blood on the cut on his head, she might have wiped his head. Since she was a Christian she probably prayed for him. She arranged his limbs and tidied up his clothing, and maybe she washed his face. Having finished these small but important tasks, she covered his body with a cloth.

About an hour afterward, a blood technician arrived and drew a sample of blood from the body. This was the Ebola blood test that Dr. Azziz had ordered a few hours earlier.

Sahr Nyokor's blood was tested twice, once by Augustine Goba and his staff, using the Harvard PCR machine, and once by Nadia Wauquier and her staff, using Wauquier's PCR machine. By late afternoon, both tests matched, and they revealed that the ambulance driver had died of Ebola. Augustine Goba sterilized and froze a half-teaspoon of Nyokor's blood serum in a plastic vial. Two days later, this blood sample, along with a large number of other samples of Ebola-infected blood, was in the air over the Atlantic Ocean, on its way to the United States for genome sequencing at the Broad Institute.

As conditions in the Ebola ward and at the hospital got worse, Lina Moses virtually stopped communicating with her husband and daughters. She wasn't the greatest communicator by email anyway, and phone connections between Kenema and New Orleans were bad, so it was hard to hear somebody talking. Moses really wanted to spare her family any vivid knowledge of the horror that was unfolding around her. Her initial expectations that the Ebola outbreak would be fairly small and manageable had not proved correct. She also didn't want her loved ones to worry about her. She felt confident that she could keep herself safe. Meanwhile, in New Orleans, school was letting out for the summer. Aron was planning to take the girls on a summer road trip along the East Coast, to see Washington, D.C., and other places and sights.

On June 20, a British epidemiologist who had arrived to help trace the disease in the population noticed that Lina Moses was running back and forth in flip-flops through a crowd of people that gathered in front of the Ebola ward every day. Some of the people in the crowd had Ebola and were vomiting. He spoke to her about the flip-flops, and he was upset. "You are insane to wear those! You must wear rubber boots!" he said. Moses felt she couldn't run in rubber boots, but she compromised by wearing a pair of old hiking boots. There were mud puddles in the dirt area in front of the Ebola ward, and as she walked through the puddles her boots got rinsed of Ebola particles, perhaps.

Moses was staying in a room at the Tulane guest house, a neglected structure with a verandah and a dead garden, on Hangha Road—a thoroughfare that runs north out of town along the base of the Kambui Hills. The Blood Diamond War had left the guest house in a state of decay, but it was comfortable and clean. It was surrounded by a high wall, and was protected by a security guard. A housekeeper named Jeneba Kanneh cooked meals and kept the house tidy. Moses's room contained a bed with a mosquito net draped over it, a bureau, a light-bulb on the ceiling, and an electric fan on the floor, pointed at the bed.

Moses had almost stopped communicating with her husband and daughters. *There's this thing called Ebola here,* she had written in an email to them, but she hadn't really explained it. She couldn't explain to them what was happening in Kenema or how she felt about it. She wore a locket around her neck that held pictures of her young daughters. Lying in bed in her room in the guest house, late at night, after another impossible-to-describe day of carrying supplies to the Ebola ward and carrying tubes full of infected blood back to the Hot Lab, Lina Moses would open up the locket on her neck and look at the faces of her daughters. Someday, would they remember her as their mother who was always gone, their missing mom? Or would they remember her as an example, a hero? All she knew was that she couldn't leave Kenema now, no matter what happened.

TULANE GUEST HOUSE, KENEMA
7 a.m., June 22

A day after the epidemiologist forced Lina Moses to get rid of her flip-flops, Jeneba Kanneh was working in the kitchen of the Tulane guest house. She had put hot water in a thermos for instant coffee and some bananas and mangos and breakfast cereal on the dining table in the parlor. She heard Moses's bedroom door bang open and Moses running into the bathroom and being violently sick. Kanneh went to the bath-room door and asked Moses if she was all right.

"It's nothing," Moses answered through the door.

After a while, Moses went back to her bedroom. But then she ran to the bathroom and got sick again. After she'd returned to bed, the housekeeper went to Moses's door and asked, again, if she was feeling all right.

"I have a touch of something."

About an hour later, Nadia Wauquier arrived at the Lassa Laboratory. When she looked into the Library—the crisis operations center—she saw that Moses wasn't there. This made her feel slightly concerned. Lina Moses had recently had a bout of malaria, but even when she'd been feverish and shaky with malaria she had kept on working in the crisis center. What would take Lina away from work? Nadia started asking around: Had anybody seen Lina?

Nobody had seen her.

Nadia sent Lina a text: *Where are you?*

A text came back: *Not feeling perfectly well*. Lina added that she would be staying home for the day.

This didn't worry Nadia, not very much. This is Lina's malaria coming back, she thought. She went across the hall to the entrance of the Hot Lab and put on PPE, and then went inside the Hot Lab and began preparing blood samples for the day's Ebola testing. Hours went by while Nadia worked in the Hot Lab. Then she de-suited and exited, and transferred her samples to her cargo-container lab and began processing them in her PCR machine.

As she worked, Nadia glanced at her phone once in a while. She was hoping to see a text from Lina with an update on her illness, but there was nothing. Lunchtime passed. Then came afternoon and there was still no text from Lina. Nadia noticed that the Sierra Leonian staffers were very aware of Lina's absence. They were obviously wondering if she was sick, and they seemed worried about her.

The afternoon dragged on. At 2:59 p.m., Nadia got a text from Lina: *I'm pretty sure I have a fever now.*

Nadia believed this was nothing serious. It had nothing to do with Lina's flip-flops, or her habit of walking into the foyer of the Ebola ward without biohazard protection, Nadia thought. Lina probably had a stomach bug. But . . . in case. Just in case. She should probably test her blood. Very quietly. If people found out that Lina . . .

Lina had seemed invulnerable, untouchable. She claimed to know where the virus was. Maybe Lina hadn't known where it was. It would be necessary to get a sample of Lina's blood. Nadia didn't know how to do a blood draw. She would need to find a blood technician who would do the draw secretly and wouldn't talk about it afterward.

Nadia exited the Hot Lab and found a blood technician named Hassan Katta. Katta was a friend of Lina Moses. He agreed to do the blood draw and keep it a secret. Working discreetly, Nadia and the tech loaded a set of PPE into the vehicle and drove to the Tulane guest house, into the walled compound, and parked behind the house, as close to the back door as possible. Some of the neighboring houses had lines of view going into the yard. If any neighbors saw a man wearing a white moon suit going into the house, it could set off a panic in the neighborhood.

Nadia and Katta opened the vehicle's doors, to further block the view. Standing between the vehicle's doors, Katta stepped into a whole-body Tyvek suit and zipped it up. He slid his feet into rubber boots, and put on a HEPA breathing mask, eye protection, and gloves. He opened a blood kit and took out a venipuncture needle and a Vacutainer red-top blood collection tube.

Nadia noticed that the needle was shaking in Katta's hands. He was extremely nervous.

They had brought only one set of PPE, so Nadia stood just outside the back door. She looked into the parlor, which was a large room with nothing in it except a dining table and some chairs. Lina's bedroom door was to the right of the parlor, and was closed. Katta entered the house, knocked on Lina's door, and went in.

He found the room flooded with sunlight and very hot. The curtains were wide open, and an electric fan sat on the floor, blowing air toward the bed. Lina was lying on the bed. She was conscious and very sick. Her face was flushed and pouring with sweat, and her shirt was saturated with dark patches of perspiration. Katta had forgotten to bring a thermometer with him. He put his gloved hand on her forehead and estimated that her temperature was 103 to 104—dangerously high. He got a tourniquet ready.

Katta's hands were shaking, and Lina Moses didn't like that. "I can pull the tourniquet," she said to him. She sat up and wrapped the strip of rubber around her arm and tightened it, and then watched as Katta tapped his gloved finger on her arm, feeling for a vein. He uncapped the needle. When the bare needle appeared, his hand started shaking again. He pushed the needle into her arm, but his hand was trembling and he missed the vein. He pulled out the needle and apologized. The needle was bloody and dancing a little. He inserted the needle a second time and again missed the vein. This triggered a bleed under the skin of Moses's arm—he'd put a hole in a vein, and a goose egg of blood rose up on her arm. Katta was very apologetic.

Finally he got a spurt of blood flowing into the tube. The tube filled quickly. He removed the needle from her arm, capped it, and put the tube into a Ziploc bag and sealed it. He exited the house and handed the bag to Nadia. She sprayed it with bleach, to disinfect the outside of the bag.

Wauquier wrapped her hand around the blood tube inside the plastic bag. The tube felt really warm, almost hot. That was a high fever. No doctors had arrived from the World Health Organization. She and Lina were isolated in Kenema, and were a long way from help. There was no need to worry, she told herself. Stay calm. Do the test.

FONNIE

LASSA WARD
Same time, June 22

Grim and unsmiling, Auntie Mbalu Fonnie stood in the foyer of the Ebola ward, next to the chart table where she often met with Humarr Khan. She wore a white nurse's outfit with a small white nurse's cap on her head. In a whispery British voice, she issued orders to her nurses and spoke with people milling around the entrance to the ward, attempting to calm panicky families who were trying to get news of a patient inside. A strong odor drifted out of the ward. Fonnie's nurses were getting overwhelmed. Some of them were staying at home, but most of them continued to work. Hospital staff members came and went, delivering messages to Auntie from elsewhere, taking messages from her to other parts of the hospital. A crowd of people seethed around the entrance of the Ebola ward, a mix of sick people and family members. In front of the ward was an enclosure made of chicken wire. People who were obviously sick with Ebola-like symptoms were supposed to stay in this area, to keep them out of contact with everybody else. Fonnie sometimes left her post and went on errands around the hospital, looking for supplies, finding Khan. Khan and she were making rounds in the general wards, looking at patients in the beds, looking for signs of Ebola infection in the patients. Auntie and Khan knew there were undiagnosed Ebola patients in the general wards, and they wanted to find them and get them out. In the general wards were still many patients lying in beds, and staff doctors were still on duty, tending a large number of patients with all sorts of illnesses. One of them was a doctor named Sahr Rogers, who would not survive the situation. Auntie had begun working fifteen-hour days at the hospital, from before sunrise to late at night.

She showed little emotion as she worked, only a seemingly desperate focus on the immediate tasks in front of her. She neither cried, nor laughed, nor smiled. She expected more from her nurses than they could give, more from herself than she could give. "It is in God's hands," she said over and over. "God holds in God."

Fonnie had recently lost her husband, Richard, and she was still grieving for him. A tall, handsome man with a wild sense of humor, Richard Fonnie had been the only person who could make Auntie laugh, or so it was said. He had built a house for their extended family, a large structure made of cement blocks, with a wall around it, on the lower slopes of the Kambui Hills. Richard died suddenly, before he could quite finish the house. Auntie had had Richard buried, as is the custom, right next to the house's foundation. This left her in charge of a large family compound. Auntie's brother, an epidemiologist named Mohamed Yillah, was living at the compound with his family, along with Mbalu and Mohamed's mother, Kadie.

Yillah, a tall, thin, quiet man, was devoted to his older sister, Auntie. He drove her to and from the hospital on his motorbike, before dawn and late at night. As they went down Hangha Road, they were an unmistakable pair—an extremely tall man with a small woman dressed in white sitting behind him, her arms wrapped around his middle. Yillah was never far from Auntie when she worked in the Ebola ward. He brought her food to eat, he ran errands and took messages around the hospital for her, and he watched over her to make sure she wasn't getting too tired.

At the Ebola ward there was a private room for the nurses, where the nurses took rest breaks. There was a table in the room that was usually piled with purses—the nurses put their purses there before they went on duty. When Auntie needed a break, a nurse would move the purses off the table, and she would get on top of the table and stretch out on it and rest for a little while. Often her brother sat in the room, watching over her as she rested on the table. She never seemed to fall asleep, though. After fifteen minutes or so she would get down from the table and go back to managing the Ebola ward.

There were now about thirty-five patients in the Ebola ward—it had been designed to hold twelve patients. The nine cubicles were jammed with cots and beds. Each cot and bed had at least two people lying on it; some beds had three people in them. Children were lying next to adults. People died in the beds next to living people. Ebola made people disoriented, and disoriented patients got out of their beds and wandered along the corridor, and fell on the floor, and couldn't get up. The Ebola nurses continued giving IV fluids to everyone who seemed dehydrated. Their biohazard suits were getting splashed with body fluids. The floors were filthy. People were dying during the night, and their bodies weren't removed the next morning. Corpses lay on beds or on the floor. When there was time, the nurses would put a body in a body bag and leave it inside a small structure next to the ward—a morgue. Ambulance crews wearing PPE would take the bodies to the Kenema cemetery, a stretch of brushy ground outside the city, which, as the city's potter's field, is full of unmarked graves. The Ebola dead could only be buried there.

TULANE GUEST HOUSE
4:30 p.m., June 22

Nadia Wauquier, standing at the back door of Lina Moses's house and holding the plastic bag containing the tube of Lina's blood, discussed the situation with Lina, who was lying in bed in her room. They decided that Lina would not leave her room for any reason, except to visit the bathroom. Nadia promised to call Lina on her cellphone whe she had gotten solid results.

Then Nadia and the blood technician drove back to the hospital, and Nadia went into the Laboratory building, holding the bagged blood tube. She stood in the anteroom of the Hot Lab and dressed herself in full PPE, and then she pushed through the glass door of the Hot Lab.

A logbook rested on a window ledge under a row of windows running along one side of the room. The logbook was the registry for all blood samples that came into the lab. Wauquier didn't want to write *Lina Moses* in the logbook—the lab staff would see her name and would be very alarmed. Instead, she wrote *Lucia Musa. Musa* is Krio for Moses. Then she removed the tube's red rubber top, inserted a pipette into the tube, and sucked up a droplet of her friend's blood, and put the drop in a very small tube, and spun the tube in a centrifuge. This made the red cells settle to the bottom of the tube, while the blood serum floated on top of the red cells. From that point she purified the serum, extracting broken RNA from the serum—the molecule that contains Ebola's genetic code. Meanwhile it grew dark.

HOT LAB
7:30 p.m.

An hour after sunset, Nadia Wauquier was in her suit and standing by the exit door of the Hot Lab, just inside. There, she stepped into a plastic tub that held bleach water, to disinfect her rubber boots, and she sprayed herself all over with bleach spray and washed her gloved hands in the liquid with special care. She sprayed the inside and outside of a box that held a number of thin glass tubes, each of which contained a purified sample of blood taken from a person suspected of having Ebola. One of the glass tubes held the sample taken from "Lucia Musa."

After having sterilized everything, Nadia took off her mask and threw it in a large plastic barrel, a biohazard barrel. She very carefully unzipped her suit and stepped out of it, peeled her gloves off her hands, and put the suit and gloves into the biohazard barrel. Now in her street clothes, and standing in the exit doorway, Nadia turned around so that her back was facing the exit and she was looking into the Hot Lab. Then she *walked backward* out of the Hot Lab. She was following protocol: You push the door open with your rear end and step backward through it. You end up in a small, closet-like room. The reason for facing backward as you exit the Hot Lab is so that you can watch to make sure you aren't accidentally dragging any contaminated material or object out of the lab, anything that might have somehow gotten stuck to your body.

Once she was out of the Hot Lab, Nadia carried the box out of the Laboratory and went around a corner into the alley where her cargo container lab was situated. She went into the lab and placed the glass tubes in the tray of her PCR machine, and started the machine running.

Two minutes passed while the PCR machine went through a cycle. Then some dots appeared on a screen. The machine went through another cycle, and more dots appeared. Every two minutes more dots came up. A pattern was beginning to form in the dots. There were several lines of dots running horizontally across the screen, and each line was getting longer as dots appeared. Each line of dots belonged to one sample of blood. If a person's blood contained Ebola virus, then a horizontal line of dots indicating that sample would begin to turn upward and start climbing. If a line of dots turned upward, this was a signal of Ebola in the person's blood.

The PCR machine would run for about an hour before it showed any reliable signals. Nadia used the time to game out her future moves: If Lina's test results signaled Ebola, she would need to be evacuated fast: Every hour counts. Lina was an American. There were no emergency evacuation plans lined up for Americans. Nobody was aware of what was happening in Kenema. Nadia herself would handle Lina's evacuation. Nadia had a boyfriend in Kenema, Hadi, who was a businessman and Lebanese. Hadi had contacts and some money. Nadia would stay with Lina no matter what. The best chance might be to get her to Switzerland.

But it might not be possible to get Lina to Switzerland. If she tested positive she would never be allowed to get on board a commercial airliner. Therefore it would be necessary to hire a private jet to get her to Switzerland. With a crew trained in biohazard. If such an aircraft couldn't be found and paid for within twenty-four hours, then Nadia planned to put her friend into an ambulance and take her to Freetown and get her admitted to a hospital there. But Ebola was starting to arrive in the Freetown hospitals. They were becoming unsafe. The medical system in Freetown could be breaking down. Nadia really wasn't worried. She watched her machine.

A PROPHET AND A VISION

NADIA WAUQUIER'S CARGO CONTAINER
10:30 p.m., June 22

Three hours later, after two tests, Nadia hadn't seen an Ebola signal in Moses's blood. She felt very calm. The virus could be in Lina's blood but at least there wasn't enough of it to trigger a signal. Nadia called Lina and told her the results. Lina was somewhat relieved. For most people, trying to imagine oneself with all the symptoms wasn't something that was easy to confront mentally.

The next morning, though, Lina was worse, sicker than the day before. She was running a high fever and was extremely weak, with vomiting and diarrhea, and with severe abdominal pain. She refused any more blood testing and went back to work in the crisis center. Her husband and daughters had absolutely no idea of her situation. After forty-eight hours, when her symptoms didn't progress, Moses concluded that she probably didn't have Ebola. But she was still very sick.

A few days later, a doctor from the World Health Organization finally arrived to bring help to Humarr Khan. He was a U.S. Navy physician named David Brett-Major. He examined Lina and gave her an antibiotic, and her symptoms abated. Afterward, Brett-Major set to work in the Ebola ward, doing what he could. He was distressed by the situation, and disturbed by what he felt were serious lapses of biosafety. The nurses were supposed to spray their hazmat suits with bleach before they took them off, but often the bleach sprayer was out of bleach, or the nurses didn't bother to spray themselves.

Then a man who worked as a nursing aide in the maternity ward tested positive for Ebola and later died in the Ebola ward. This was the second member of the Kenema medical staff to die of Ebola—the first victim had been the ambulance driver Sahr Nyokor. The maternity aide's death terrified the nurses who were still working. If somebody working in the maternity ward could catch the virus, anybody on the staff could catch it anywhere in the hospital.

As news of the aide's death spread, a teenage boy suddenly walked through the hospital gates and began shouting. He was a wiry stick of a kid, about eighteen, and his name was Wahab. Wahab was known around Kenema. He was an herbalist who treated sick people with preparations made from plants, and he was considered to be a visioner who could see into the future. There were people in town who thought that Wahab was a little crazy, while others believed that Satan had gotten into his mind and made him confused. Still others believed that Wahab really could see the future, and that his visions were significant and powerful. Wahab paid house calls. If you wanted to know your future, he would show up at your door, if he chose to. He didn't charge money for his visions, he gave them for free, and only if visions came to him.

According to Wahab the Visioner, fate was real but it wasn't absolute: One's fate could sometimes be changed. If you were fated to have something happen to you, Wahab could see your fate hanging in the future, but he also sometimes saw a way you could change your fate.

Now, he began walking around on the paths outside the hospital buildings, and he lingered at the maternity ward. "Oh!" he shouted in a piercing voice that carried into the maternity ward. "Oh! A nurse has died here!" (He was speaking in Krio—his words are an English translation provided by a member of the hospital staff who heard Wahab speak.)

"Three nurses will die!" Wahab shouted. "One nurse has already died! Two more will die! Three nurses will die! This cannot be changed!"

This was the unchangeable future he saw, that three nurses were fated to die. One nurse was already dead, and two more nurses, the fated ones, were doomed, though Wahab didn't say who they were. Wahab began walking up and down the walkways that joined the wards together, circling the wards and speaking his vision in a wild voice, saying that two more nurses were going to die. His voice carried into the wards and was heard by nurses and patients.

Then Wahab offered the hospital his vision of how to change fate. "All the living nurses at the hospital must do sacrifice or prayer!" he cried. "You must do a ceremony!" If the nurses failed to do a ceremony as he recommended, then many more nurses were going to die, many more nurses than just three would die. But no matter what the Kenema nurses did in a ceremony, three of them were fated to die. "Even if you can make that ceremony or you don't do it, three nurses are going to die," he shouted. "But if you don't do that, if you don't make the ceremony, many nurses will die! You must do a sacrifice and prayer!" Then, as suddenly as he had appeared, the boy slipped out through the hospital gates and disappeared into the city.

Wahab's vision frightened the nurses who heard him speaking— those who were still working at the hospital. His word of prophecy went around the hospital quickly, and it went around the nurses' families in the city. Wahab had been vague about certain details, however. He hadn't explained what he meant by a "ceremony" or a "sacrifice." It wasn't clear exactly what kind of a ceremony Wahab wanted the nurses to perform in order to prevent many more nurses from dying. And he hadn't identified the two nurses who were doomed no matter what, the ones who couldn't escape their fates. The nurses wondered who among them were the two who couldn't escape death.

The next morning—a Friday—a crowd of nurses gathered in a

dusty, open area near the maternity ward, where the dead maternity aide had worked. They offered Christian and Muslim prayers, and sang hymns, and they offered penance to God, asking God to forgive them for their sins and to spare them and the hospital from more deaths.

Lucy May, the pregnant nurse who had taken care of the ambulance driver as he died, may have been in the crowd that morning. She may have sung hymns with the nurses, since she had a fine voice and sang in the cathedral choir. Eight days earlier, she had wiped blood from Sahr Nyokor's head just before he died of Ebola. Since she was a night nurse at the Annexe ward, Lucy May would have just gotten off her shift on the morning of the "sacrifice." But she was close to the end of her pregnancy, and since she was expecting a baby soon, she may have skipped the ceremony and gone home right after her shift ended, and she may not have been feeling very well.

At eight o'clock that evening, Lucy May didn't report to work in the Annexe ward for the night shift, which was her regular shift. Her colleagues assumed that her pregnancy was bothering her. The next day, Saturday, she stayed at home in bed all day, and by then she wasn't well at all. On Sunday morning she didn't attend mass or sing in the choir at St. Paul's Cathedral. By Sunday night Lucy May was desperately ill. Somebody in her family called the hospital for an ambulance, and she was admitted to the Annexe ward, her place of work.

She was put in a semi-private room, and Dr. Humarr Khan examined her. Because she was a member of the hospital staff, Khan became her attending physician and paid close attention to her. By now it was clear to Khan that she might have symptoms of Ebola, and he ordered a blood draw so that she might be tested. Even if it was suspected she had Ebola, she had to remain in the semi-private room in the Annexe ward. It would have been unethical to put a pregnant woman in the Ebola ward without a blood test proving she had Ebola. If she wasn't infected but they put her in the Ebola ward, she and her baby would certainly catch the virus.

Ebola virus is known to be nearly one hundred percent fatal for pregnant women and their unborn babies. The virus typically kills the baby inside the womb, and triggers flooding hemorrhages from the

birth canal as the baby is being born. The baby, itself infected with
Ebola, is either stillborn or dies shortly after birth. This aspect of Ebola
disease has been known to doctors since 1976, when Sister Beata, the
midwife of Yambuku, delivered at least two babies during hemorrhagic
childbirths. It has also been known since the death of Sister Beata following those deliveries that Ebola virus is extremely dangerous for
medical caregivers who come into contact with blood or body fluids of
an Ebola-infected pregnant woman during childbirth.

A sample of Nurse Lucy's blood was sent to the Hot Lab to be
tested. By the next day, Nadia Wauquier and Augustine Goba had
confirmed that Lucy May had Ebola. Her blood showed a high
positive—she already had a heavy concentration of virus particles in
her bloodstream. It meant that her and her baby's chances of death
were close to one hundred percent. Lucy May was transferred to the
Ebola ward, where she came under the care of Auntie Mbalu Fonnie.

RESCUING LUCY MAY

July 3

When Lucy May was admitted to the Ebola ward, Mbalu Fonnie, Humarr Khan, and the Ebola nurses decided that she should not be placed among the other patients. The ward was a horror. They found a small space for Nurse Lucy in the little nook at the end of the corridor in the red zone. They moved a cot into the nook and put Lucy on it, in order to give her some privacy and to spare her from seeing the ward. At night, a senior Ebola nurse named Alex Moigboi took over Lucy's care. He was working twelve-hour night shifts in the Ebola ward while wearing PPE, tending around thirty Ebola patients. Nurse Alex did what he could for Lucy, visiting her often and tending her needs, and trying to give her some company so that she wouldn't feel alone. He set up an IV for her and gave her a saline drip, to make sure she didn't get dehydrated.

Lassa and Ebola produce similar effects in pregnant women: profuse hemorrhaging, death of the baby, death of the mother. Both viruses are nearly always lethal to the baby and typically lethal to the mother. Nevertheless, over the years, Auntie Mbalu Fonnie had rescued a number of hemorrhaging pregnant mothers who seemed doomed. Auntie's rescue technique was to remove the fetus from the mother as quickly as possible, by abortion or induced delivery, and then she would give the mother a dilation and curettage. Often the virus killed the fetus and triggered a miscarriage, though at times a Lassa-infected baby would be born alive, though it rarely survived for very long. After the baby was removed through abortion or was born naturally, the D&C procedure was used to scrape out any remaining placental or fetal tissue from the inner wall of the uterus.

For reasons that aren't clear, the Lassa rescue procedure seems to increase a Lassa-infected pregnant mother's chances of survival dramatically—giving her perhaps as much as a fifty-fifty chance at survival. Auntie wondered if the Lassa rescue procedure would work with pregnant woman infected with Ebola. Could the procedure save Lucy May? Auntie had reason to think it might work. In fact, she had already seen an Ebola-infected pregnant woman survive her ordeal, even though the evidence predicted she would die. The survivor was Victoria Yilliah, twenty, the first confirmed Ebola patient discovered at the Kenema hospital. Ms. Yilliah had lost a stillborn baby in a hemor-rhagic miscarriage and had ended up in the Lassa ward under Auntie's care. Auntie, mistakenly thinking that Ms. Yilliah had Lassa, had done a version of the Lassa rescue. Ms. Yilliah had already lost her baby, but Auntie had given her a dilation and curettage. Ms. Yilliah had been hemorrhaging at the time—and she had survived.

It seems crazy to scrape a woman's uterus with a scalpel when she has Ebola and is hemorrhaging, but Ms. Yilliah gradually recovered after her D&C. In thinking about the previous experience with Victo-ria Yilliah, Auntie started considering trying to save Lucy May with a Lassa rescue. The question was timing. If she was going to try a res-cue, when should she make the attempt? Lucy was at a late stage of her pregnancy. The baby could be born naturally, and it might have chance of survival. Auntie decided to wait, not make a decision yet. She would not abort Lucy May's baby unless it died in the womb. She would let God make any decision to end the baby's life. Fonnie monitored Lucy and her baby, directing the nurses at their tasks and sometimes putting on PPE and tending Lucy herself. At night, Nurse Alex Moigboi tended Lucy. Her face settled into an expressionless mask.

8 p.m., July 3

A nurse exited from the red zone and told Mbalu Fonnie that Lucy May was having a miscarriage. Fonnie dressed herself in PPE and entered the ward; she then went to Lucy's bedside in the little nook at the back of the ward. She listened to Lucy's abdomen and confirmed that her baby was dead.

Bleeding from the birth canal had begun. This indicated that Lucy had gone into disseminated intravascular coagulation, or DIC. Tiny clots had appeared in her bloodstream, and they had lodged everywhere in small vessels, and her blood had lost its ability to clot, and was running out. The bleeding could be coming from the walls of the uterus or the placenta.

Fonnie called three nurses to the ward and informed them that Lucy's baby was gone. She asked the nurses if they would join her in an attempt to save Lucy's life. Fonnie was going to induce childbirth and try to get the deceased fetus out of its mother as quickly as possible. The procedure had to be done immediately. Lucy May was bleeding, and she could die at any time.

The three nurses were absolutely terrified, but they agreed to help Fonnie because Lucy was a fellow nurse. The three nurses suited up in the cargo container. An extreme shortage of disposable biohazard suits had developed. The nurses were reusing their suits, spraying them with bleach, or not spraying them, then taking them off and putting them back on the next time they went into the red zone.

After the nurses had suited up, they came to Lucy's bedside. One of the nurses was Princess Gborie, the nurse who had tended to ambulance driver Sahr Nyokor just after he had died. Auntie had recently called Princess Gborie into the Ebola ward and trained her in the wearing of PPE. Princess was not experienced, and had only worn bio-hazmat gear for a few days. The other two nurses were women named Sia Mabay and Fatima Kamara. They had suited up and gone into the red zone because Lucy May was their friend and colleague. The three nurses were beyond frightened.

At about 8:10 p.m. Fonnie instructed the nurses to set up an IV drip of Pitocin, a drug that induces birth contractions. Lucy was conscious and in extreme pain. There was no anesthetic available. The ward had run out of many basic supplies.

The drug was infused into her quickly. Auntie wanted to get the baby out as fast as possible, because for every minute the baby stayed inside Lucy her risk of death went up. Soon Lucy's contractions began. The pain must have been unimaginable and unendurable, since the contractions of childbirth were occurring in tissues that were saturated with Ebola particles and packed with blood vessels, and the vessels were breaking down and leaking large amounts of blood. She was hemorrhaging from the birth canal. Her blood wouldn't coagulate. The bleeding increased as labor began. There was blood in the cot. They held her knees, talked to her, swabbed her with towels or pads. They held her hands. They spoke to God. Lucy went into a luminous bleedout in the cot. The nurses' gloves got bloody. There was blood on the sleeves of their suits. They tried to keep themselves calm for the sake of Lucy. There were no blood supplies in the ward. There was no possibility of giving her a blood transfusion.

Fonnie ran her hand through the birth canal to the cervix and felt the cervix and estimated its dilation. Since the baby was no longer alive, she would remove it quickly in a breech birth position, feet first. She wouldn't wait for the cervix to dilate fully, she would pull the baby out as soon as possible. She would not use sharp instruments to cut into the deceased baby or try to break up the tissues of the baby, since there would be a risk that Lucy herself could be cut by an instrument, and Lucy couldn't afford an injury like that. As Auntie worked, she got blood on her gloves and on the sleeves of her protective suit. Pinpoints of splashed blood may have appeared on her transparent surgical face shield, and on her breathing mask, and blood drops may have landed on a small area of bare, exposed skin on her throat below the HEPA mask that covered her nose and mouth.

Auntie got her hand through the cervix and pulled the baby out. It emerged in a gush of fluid and blood. Fonnie pulled out the placenta, which seemed to be the main source of hemorrhaging, and she cut the umbilical cord away from the deceased child.

At the sight of the stillborn child and the blood and fluids in the cot and all over themselves, the three nurses would have received visual confirmation of their fears that they themselves would get infected by this procedure. There was too much fluid and blood, the virus was everywhere, all over them, all over Lucy, coming out along with the child.

We cannot know what went through Auntie Mbalu Fonnie's mind when she saw Lucy's child, or when she perceived how wet her suit was. Maybe Auntie was too tired to think much about herself or what could happen to her. She was a widow, she missed her husband, her ward had been destroyed by Ebola. It seems possible that she thought to herself, or whispered, "God will drive this."

We don't know if Auntie proceeded with a D&C afterward. She may have decided it was too risky to scrape the uterus with a scalpel if there was so much bleeding. After the procedure was finished, the nurses and Auntie cleaned Lucy, and prayed for her, and she rested quietly. At around nine o'clock that night her breathing became labored. There was no oxygen in the ward, no way to assist her breathing. At 9:15 p.m. Lucy lost arterial blood pressure and went into shock. Her blood pressure collapsed, her heartbeat became irregular, and at about 9:30 p.m. she had a cardiac arrest. The rescue of Lucy May had failed.

When it was clear that she had passed, the three attending nurses burst into screams, and their screams turned into roars of agony. The sounds came out of the Ebola ward and drifted across the hospital grounds in the darkness, frightening people all over the hospital. Fifty yards down the hill from the Ebola ward, Nadia Wauquier was working inside the Hot Lab in a moon suit. She heard the nurses' cries and knew that something terrible had happened in the Ebola ward.

We can imagine that Auntie might have been somewhat stern with Princess Gborie, Sia Mabay, and Fatima Kamara. She might have asked the three nurses to quiet themselves. We can also imagine that she might have told them that this was God's will. We can easily suppose that Auntie would have wept, too, joining her nurses' tears.

Eventually Auntie went to the decon area and removed her biohazmat suit, which was smeared with birth fluids and blood. There was a bleach sprayer in the decon area. No one knows if it contained any bleach that night. The sprayer was often empty or not used, and a shortage of bleach had developed at the hospital. No one knows whether Auntie sprayed her suit before she took it off. Underneath she was wearing her usual starched white outfit, which was now completely soaked with sweat. Her brother, Mohamed Yillah, was waiting for her. He started his motorbike, and she climbed onto the seat behind him and wrapped her arms around him, and he drove her home. He caught it, too.

The next afternoon, Wahab the Visioner walked through the hospital gates and began shouting. He knew all about the death of Lucy May.

CANDLE FLAME

Afternoon, July 4

Wahab the Visioner walked swiftly around the hospital grounds, agitated and shouting, driven by a vision of the future that seemed to torment him like stinging flies. "Oh!" he cried in a piercing voice. "Oh! You did not make enough sacrifice! You did not make enough sacrifice!" The nurses' ceremony had failed, he shouted. Lucy May had been one of the three nurses who had been fated to die. Now *many more* nurses were going to die. But there was still a chance that these many doomed nurses could save themselves. The future could still be changed. Wahab's voice carried into the main wards of the hospital, and he may have gone inside the wards and prophesied directly at nurses and patients—those who were still left at the hospital. "For now," he shouted, "you must sacrifice a candle. All living nurses must make this sacrifice. You must light candles all around the hospital compound. All the nurses must do this! If you don't do this sacrifice a lot of nurses are going to die. Even an important doctor will die." Abruptly the young man was gone.

After sunset, around fifty nurses gathered in front of the maternity ward. Many of them were wearing civilian clothes, because they had stopped working at the hospital. They lit candles and began walking quietly around the hospital grounds, singing gospel songs in Krio, in three-part harmony. Their voices were serene and gentle. More and more medical staff joined the crowd as it flowed around the hospital, and each person held a burning candle in hopes that the candle would be a suitable sacrifice, and that the holder of the candle would be spared. As nurses sang, they asked God to spare them and the hospital from more deaths.

"It was shocking, it was beautiful, and it was terribly, terribly sad," Nadia Wauquier remembered.

The crowd grew to more than a hundred members of the medical staff, and eventually it stopped in front of the children's ward. The children's ward has a large open area covered by a protective roof, where parents and their children gather while they are waiting to see a doctor. The roof protects them from rain and sun. The crowd gathered under this roof in the waiting area and in the open air around the waiting area. The children's waiting area and the walls of the children's ward glowed with candlelight as the nurses sang. Wahab's vision seemed to hang in the air in the darkness above the many small flickering glows of the candles and the faces of the medical staff. Wahab had predicted that if the sacrifice of their candles didn't prove to be enough, many more would die, among them an important doctor would be among the dead. If Auntie Mbalu Fonnie couldn't stop the virus no matter what risks she took, and if candle flames didn't stop the virus, then there would be no way to stop it, no way at all.

PART THREE

THE ANCIENT RULE

JOURNEY TO KINSHASA

In order to have a clearer view of emerging viruses, we need to look back in time to the crisis of 1976, when Ebola first manifested itself at a remote Catholic mission in the lowland rain forest of north-central Zaire. If we look closely at this outbreak, the first outbreak of Ebola, we can learn things that may help us prepare for the next emerging virus, whatever it is and whenever it comes. It seems virtually certain that another Level 4 emerging virus, one that is perhaps far more contagious than Ebola, will cross from the virosphere into a person somewhere on earth. That person will pass the agent to someone else, and quickly the agent will move along airline routes, traveling inside air passengers, and it may be able to ignite spreading chains of infections in cities, as Ebola did. Since there will be no vaccine and no treatment for such an emerging virus, and because it may spread easily from person to person, the virus will be terrifying and will seem virtually unstoppable.

If we are going to stop the next virus, it pays to study history. There is much to be learned by looking at the people who engaged with Ebola during its first known encounter with the human species. We can study their lives and deaths, and we can watch what they did as they confronted the unknown. We can watch the moves they made as they engaged with the virus—and the moves the virus made in response. We can learn their secrets.

Ebola is an entity of a kind, though it is not a conscious thing. It was not even a thing, it is uncounted numbers of things, each one striving, in a biological sense, to survive and replicate. A growing swarm of Ebola particles has no awareness of itself as an entity. It has no memory of the past or ability to anticipate the future. The swarm has no emotions, no desires, no fear, no love, no hate, no pity, no plans, no such thing as hope. Yet, like all forms of life, each particle in the swarm possesses an unalterable biological will to copy itself and perpetuate its genetic code through time.

As we look at the crisis of 1976, we can see how something that seemed unstoppable was stopped, and we can see the spiritual and human cost to the people who faced Ebola and tried to stop it. If we can learn something from the events of 1976, it can sharpen our perceptions of the battle that was unfolding at Kenema Government Hospital in 2014 and, indeed, was expanding into the world. We can learn something about human character in moments of life-or-death crisis, and we can observe the dramatic interaction between the human species and nature, which has always been a subject of deep interest to this writer. The simple question faced by those who encountered the virus was how to kill it before it killed them.

SOMEWHERE IN THE AIR ABOVE THE CONGO RIVER, ÉQUATEUR PROVINCE, ZAIRE
About 9 p.m., September 26, 1976

Jean-Jacques Muyembé, MD, PhD, sat in a seat in a Fokker Friendship passenger aircraft, listening to the whining hum of the plane's twin turboprops. It was dark outside the window. The plane was flying over a region of nearly unbroken rain forest, tracing the Congo River along its course upstream. There were no lights visible down below—no electricity and no towns, although there were villages here and there, nestled in small breaks in the forest canopy or hidden under the canopy. Some of the villages were inhabited by people who spoke Bantu languages, while other villages, and small camps, were occupied by Twa people, who are of very short stature, and are an ancient people who have lived in the central African rain forest for tens of thousands of years, far longer than any other group has lived there.

Muyembé was anxious to deliver his sick patients, Sister Myriam and Father Sleghers, to the best hospital in Kinshasa. Their illness, a polymorphic disease with puzzling features, was expected to progress rapidly and potentially be fatal. Muyembé also wanted to bring his samples of blood and liver back to his lab before they rotted in the tropical heat. By analyzing the samples in his lab, he hoped to identify the disease, and then, perhaps, find a way to stop it. He strongly suspected it was typhoid fever or yellow fever.

After a while, a cluster of lights appeared, and the Friendship descended and landed at Kisangani, a city on the Congo River that was once known as Stanley Falls.

The air terminal at the Kisangani Airport was a decay-in-progress made of rain-spalled concrete, guarded by soldiers loyal to the maximum president of Zaire, Mobutu Sese Seko. The doctors helped the church people get settled in chairs in the waiting room of the terminal, and then bought sodas for them. Sister Myriam felt well enough to drink a Fanta or a Coca-Cola. The little group passed much of the night—their second night since leaving the Yambuku mission—sitting in the waiting room at the Kisangani Airport. J. J. Muyembé kept the box of samples next to him on the floor, where they were as warm as Zaire. He knew the blood samples and the piece of liver were decomposing, but he remained hopeful that something would be preserved.

In the early hours of the morning, a Boeing jet flown by Air Zaire landed at the Kisangani Airport. The doctors and their patients got on board, and the jet carried them westward over lightless tracts of rain forest dissected by pythonic rivers. Sometime after dawn, the rain forest gave way to tracts of savanna and gallery forest and cultivated land, and Kinshasa appeared, a brown sprawl giving off haze. The plane landed at N'djili International Airport.

They got off the plane. By then, the sun had risen and another day had begun in the capital. Muyembé escorted the nuns and the priest out of the terminal to the taxi area. The air was thick with diesel fumes and smoke from cooking fires, and there was an ever-present chatter of motorbikes. He got them a taxi and told the driver to take the nuns and the priest to Ngaliema Hospital. He bid them goodbye, assuring them that he'd contact them as soon as he had any information about the disease. The piece of liver had been in the heat for two days. He was very anxious to get it into a microscope and look at it.

A taxi dropped Muyembé at the university. It had a comfortable-looking campus spread with modern buildings. He hurried into his lab with the samples. Muyembé had a lab staff, and working with them, he divided the piece of liver into several pieces. He and the staff then prepared some very thin slices of the liver and mounted them on glass slides. Muyembé wanted to get multiple opinions on these liver samples. Could this be yellow fever? If not, what was it? He had two different colleagues look at the slides and make findings. He put a slide holding a very thin slice of the liver into a powerful microscope. He needed to look, too.

Under high magnification, liver tissue infected with yellow fever virus would show distinct changes. But when Muyembé looked at the tissue he saw nothing definite. There was nothing to see. The tissue had rotted into a mush. This was perfectly unsatisfactory. He couldn't rule out yellow fever or rule it in.

He still had the blood samples, though. With these he would be looking for typhoid. If the disease at Yambuku was typhoid, then the blood would be teeming with typhoid bacteria. The bacteria would have multiplied in the warm, rotting blood. He and his lab staff set out some petri dishes and put drops of the blood on them, and then stored the dishes in a warm place. It would take a day or two for any colonies of typhoid bacteria to grow on the petri dishes. If he saw any typhoid colonies on the dishes, then he would know that typhoid fever was raging at Yambuku.

NGALIEMA HOSPITAL, KINSHASA
Midday, September 27

Ngaliema Hospital sits on a hill overlooking the Congo River at the lower expanse of the great Malebo Pool, a broad, sluggish detention of the river before it pours into the Great Falls of the Congo. The Ngaliema hospital is a neat arrangement of low pavilions, painted white and set around rectangular courtyards covered with grass. Sister Myriam was put in a private room in one of the pavilions, where she got sicker. She began vomiting and had episodes of diarrhea. Sister Edmonda cared for her, but she didn't wear rubber gloves or a protective gown or mask.

Father Sleghers, the feverish father superior, got lucky. It turned out that, indeed, he did have malaria. There was nothing else in his blood except malarial parasites, and malaria medicine eventually made him better. But Sister Myriam did not have malaria, and she declined rapidly at the hospital.

J. J. Muyembé had been tracking Sister Myriam's illness, and at the same time he had been watching the petri dishes into which he'd placed a few drops of the blood he'd collected from various people who'd been exhibiting the symptoms. Nothing yet had grown on the dishes. At this point he began to think that he might not have a good handle on the disease. It might *not* be transmitted through bites of mosquitoes or through oral consumption of contaminated food and liquids. It might be contagious, in fact. He phoned Sister Myriam's attending physician at Ngaliema Hospital. "We don't know the exact kind of disease this is," he said to him. "We must take care, and we must be prudent." He advised that the hospital staff take basic infectious-disease precautions with Sister Myriam. They should consider her to be potentially contagious.

"Ce n'est pas un problème," the doctor replied—it's not a problem. "I think it's a simple typhoid fever."

Muyembé was extremely busy with his duties at the university. In addition to running the microbiology lab, he was the dean of the medical school. While he was waiting for results from the petri dishes, he met with professors and students in his office and around the campus. At this time Zaire was a newly independent country, and there was a sense of optimism and energy in the air. The campus was a lively place, and Muyembé got caught up in meetings and work.

The day after he returned from his trip to Yambuku, however, he got a grim piece of news. Father Germain, the thin, older, goateed curate of the mission, who had given last rites to Sister Beata, had fallen ill. This was very disturbing. Whatever it was, it was spreading.

The next day, Sister Myriam started bleeding. Muyembé considered the advice he'd just given to the doctor at the hospital, to consider that it might be contagious. Meanwhile no colonies of typhoid bacteria had grown on the petri dishes. Therefore the disease wasn't typhoid.

At this point he was at a loss. He also began wondering just slightly about Sister Myriam and himself. That rash he'd seen on her torso and breasts. With the red bumps and the petechiae, the blood spots. He had noticed how rapidly it spread up her neck and down her arms. He began to wonder if he had been exposed to something.

Sister Myriam's bleeding got worse, and it became frankly hemorrhagic, and the rash on her body darkened and became bruise-like, and her eyes turned bright red. She bled from her gums and her intestines. They began giving her blood transfusions to replace the blood she was losing from the orifices of her body. They were pouring blood into her and the blood was coming out through her intestines. Sister Myriam's care companion, Sister Edmonda, couldn't manage the workload of caring for Sister Myriam, so a hospital nurse named Mayinga N'Seka was assigned to care for Sister Myriam as well. Nurse Mayinga was twenty-three, and she had come to Kinshasa from her village near the city to work as a nurse. Nurse Mayinga and Sister Edmonda were dealing with a lot of blood that was coming out of Sister Myriam.

Muyembé thought about what was happening in that hospital room. He also thought about Father Germain, whom he'd met at the dinner table, and who might be dying in Yambuku. He remembered the long, hot ride in a crowded Land Rover, when he'd been pressed up against Sister Myriam. He could feel the moisture of sweat from her skin touching him, her arm bumping against his arm. Her skin had been pale; the rash, the mottling, had been easy to see in her skin. It would not be so easy to see the rash on an African person.

Then word came that Sister Myriam had died. J. J. Muyembé didn't know what had caused her disease, but it seemed to be a virus. A virus without a name. He thought about his family. He had a wife and children. An infection by a virus has an incubation time, which is the time between the moment the person is infected and when the person begins to develop symptoms. During the incubation period there are no symptoms. Meanwhile you feel nothing. Muyembé wondered if he was in the incubation period of a virus. He felt all right.

Another nun at Yambuku, named Sister Romana, had also fallen ill. Sister Romana quickly died in the women's ward at the Yambuku hospital, and several hours later Father Germain died in the otherwise empty men's ward. Then Sister Edmonda—Sister Myriam's traveling companion, who'd cared for her at Ngaliema Hospital—got symptoms. Sister Edmonda's case was not as severe as Sister Myriam's, but she developed black diarrhea. Nurse Mayinga, having cared for Sister Myriam, now gave care to Sister Edmonda. In the early hours of the morning on October 14, Sister Edmonda died in her room at Ngaliema Hospital.

KINSHASA
October 15

About thirty hours after Sister Edmonda died, Nurse Mayinga woke up in the morning and realized that she had a fever. This frightened her deeply. Rather than reporting to work at Ngaliema Hospital that morning, she took the day off, and went all over Kinshasa seeking medical help—she didn't want to tell the doctors at Ngaliema Hospital that she might have caught the nuns' disease. Running a fever, she spent hours waiting at the emergency room of the city's biggest hospital, called Mama Yemo, hoping that a doctor would see her. During this time she was in contact with many other people—Mamo Yemo Hospital had a huge, busy waiting room. She couldn't get a doctor to see her, so she went to another hospital. Finally Nurse Mayinga returned to Ngaliema Hospital and told the doctors she was sick. The doctors isolated her in a room. News reports had begun to appear about the Yambuku disease, and when news came out on the radio and in newspapers that Nurse Mayinga had gone all over the city while she had the Yambuku disease, the city went into a panic. Nurse Mayinga may have spread the Yambuku disease into Kinshasa. The disease had reached the capital.

As dean of the medical school, J. J. Muyembé took responsibility for tracing Nurse Mayinga's contacts during the time when she had gone around the city seeking medical help while she was running a fever. Muyembé and the investigators discovered that Nurse Mayinga had had face-to-face contact with at least two hundred people in Kinshasa—in the space of just a few hours. All of them had to be found by surveillance workers and watched, since the infectious agent could appear in any of them. Muyembé could not forget that he himself had been in close contact with Sister Myriam. Mayinga had caught the virus from either Sister Myriam or Sister Edmonda, or from both of them.

Muyembé began seeing in his mind's eye flashes of certain incidents while he'd been investigating the disease at Yambuku. He could see, and feel, the cadaveral blood running over his fingers and dripping from his wrist. His exposures to the virus had been massive. It could be growing in him now.

He again thought about his wife and children. He thought about the close contacts he'd had with many people at the university. Faculty members, students, lab technicians, citizens. He began taking his temperature twice a day, in the morning and evening. He couldn't bear living at home with his family, he might give them the virus, and so he moved out of his house and began sleeping in a room at the university. As the days went by, he kept visualizing all the terrible symptoms of the disease, including that strange rash, with the red pimples and leaks of blood under the skin. It had been obvious on Sister Myriam's white skin. But had he actually seen that same rash on the dead nurses, too? It would have been less visible on their darker skin, but he thought he *had* seen a rash on their bodies. He wondered if his temperature had risen slightly. He inspected his skin, wondering if he would start to see little leaks of blood appearing under the surface of his flesh, and spreading.

BUSH DOCTOR

While Sister Myriam lay dying at Ngaliema Hospital, a Reverend Sister of the Catholic Church got in touch with a doctor in Kinshasa named Jean-François Ruppol and asked him for help investigating the disease. Dr. Ruppol, who was then thirty-eight, was the director of the Belgian government's medical aid mission in Zaire, called the Fonds Médical Tropical, or Fometro. Ruppol was a small man, with a sharp chin, blue-green eyes, a leathery face with a tropical tan, and wavy light brown hair, and he was reputed to be a man with a sharp temper. Ruppol lived in a whitewashed stucco house on Avenue Mfumu Lutunu in the downtown with this wife, Josiane Wissocq, and two young daughters. As head of the Belgian medical mission, he was in charge of about two hundred doctors working across Zaire.

Ruppol traveled frequently across Zaire as he managed the doctors under his supervision, visiting small, rural hospitals, treating patients, giving advice, and helping the hospital staff do their work. Whenever he arrived at a rural hospital, the first thing Ruppol did was to organize the hospital's dispensary. Then he would start seeing patients. Meanwhile, word would spread through the local villages that an important doctor had arrived, and patients would start coming into the hospital for treatment. They came from up to fifty miles away. They either walked to the hospital or were carried on special Congolese travel chairs by family members. Ruppol treated the patients with whatever medicines and supplies he had on hand, doing everything from dispensing worm medicine to occasionally delivering babies. Jean-François Ruppol was what is known as a bush doctor.

As a part of his work as a bush doctor, Ruppol was an epidemiologist. He tracked outbreaks of sleeping sickness, kept statistics, and tried to stop the outbreaks. Sleeping sickness is a lethal, difficult-to-cure disease that is spread through bites of the tsetse fly. Sleeping sickness could devastate a village. When it got into a village, so many people would die that the survivors would sometimes abandon the village and move elsewhere.

Ruppol agreed to help with the Yambuku mystery, and he got written orders from the government of Zaire to try to identify the agent and arrest its spread. He was joined on the mission by a French army doctor named Gilbert Raffier, and by an energetic Congolese physician from Kinshasa's Mama Yemo Hospital, Dr. Buassa Krubwa. (Like many Congolese doctors, he went by his first name, as Dr. Buassa.)

Ruppol wanted to collect some blood samples at Yambuku. He'd been in touch with J. J. Muyembé and knew Muyembé hadn't been able to keep his samples cold, and they'd rotted. Ruppol went to a beer brewery in the city and rented several large cylinders of compressed carbon dioxide gas. The gas could be used for making dry ice, which he hoped would keep his samples of blood really cold.

Doctors Ruppol, Raffier, and Buassa arrived at the military airfield in Kinshasa just after dawn on October 4. A C-130 Hercules military transport aircraft of the Zairian Air Force was sitting on the tarmac waiting for them. But when they tried to load their gas tanks of CO_2 onto the plane, the pilot told them to forget it. The plane had reached maximum takeoff weight, he claimed.

The plane was bound for a town in the north where President Mobutu was building a palace. Ruppol and his colleagues looked around the plane's cargo hold and discovered it was packed with a number of items destined for the palace. There were crates of imported vegetables, crates of local vegetables, Belgian beer, cases of wine from Burgundy and the Médoc, cases of Champagne, Parma hams, tins of pâté de foie gras, two crates of Camembert cheese from Normandy, and a large number of concrete blocks. Everything was for the President's palace. The pilot said that if the doctors' gas cylinders were loaded onto the plane, it might not get off the ground. The Hercules could turn into a delicacy-enriched fireball at the end of the runway.

Ruppol began what he described in his journal as "difficult negotiations" with the pilot. Bribes were common in Zaire, but Ruppol never offered bribes, never. That way led to perdition for a medical man. Instead, Ruppol offered the pilot a cigarette and began a tedious discussion. The pilot kept refusing. Ruppol's spleen rose. He had friends on the cabinet. He would call in several cabinet ministers and they, not he, would settle this *affaire*. When his mention of cabinet ministers didn't impress the pilot, he began dropping the president's name. President Mobutu will want to hear about this. Finally Ruppol said he was going to call the president. The pilot obviously didn't believe him.

So Ruppol went to a phone at the airfield and called President Mobutu. After a short wait, Ruppol was told that the president wasn't available to speak with him. Ruppol walked back to the Hercules and told the pilot that the gas cylinders must be brought on board immediately.

The pilot finally said that he would take the tanks, but, for safety, something of equal weight would have to be left behind. For some reason the cement blocks were not eligible for jettison. The pilot looked around the cargo hold briefly, made a quick decision, and ordered that the two crates of French Camembert cheese be offloaded and left on the tarmac. The tanks of CO_2 were brought on board. After a brace-for-impact kind of takeoff, the Hercules lumbered into the air.

Years later, the French Army doctor Gilbert Raffier recalled in his memoir, *Africa from A to Z*, that as the Hercules took off he experienced a sharp pang of regret as he thought about all that beautiful cheese from Normandy sitting on the tarmac and spoiling in the tropical sun. As a Frenchman, he had a painful grasp of the full dimensions of the tragedy that had just occurred. This was a sacrifice, but it had to be done in order to keep the blood samples cold. The plane flew for hours and landed near the president's palace. The doctors slept in a hostel in town. The next morning a military helicopter delivered them and the cylinders of compressed CO_2 gas to Bumba Ville.

When Ruppol and Raffier stepped off the helicopter, they found the town in a panic. The national government had thrown a quarantine around Bumba Zone, and soldiers had set up roadblocks and weren't allowing anybody to leave. After meeting with the district commissioner, the doctors walked around the town putting up notices for a meeting to be held in the marketplace that day. The notices said that at noon there would be doctors at the marketplace who would explain the disease and make recommendations. Then Ruppol and Raffier went to the Bumba hospital to see if there were any patients afflicted with the mysterious disease whom they could examine, while Dr. Buassa met with the district commissioner to coordinate a government response to the outbreak.

When Ruppol and Raffier arrived at the Bumba hospital, they discovered that there were only two patients in the hospital with the disease. They were a husband and wife, and they had been isolated in a room. The man was the headmaster of the Yambuku boarding school, and the wife was a teacher and administrator at the school. They had caught the virus from their six-month-old baby, which had died.

Ruppol and Raffier paused outside the door of the couple's hospital room. Clearly there was something infective in the room. They hadn't brought any bioprotective safety supplies with them except for rubber gloves. They put on the gloves and entered the room cautiously.

The woman was lying curled up in bed, on a plastic mattress pooled with urine, begging for help. Her husband was sitting on a lounge chair facing her. One of his legs was bent, with the foot planted on the floor, while the other leg was extended straight out. One arm was bent at a right angle and propped on the arm rest, and the hand of that arm was clenched into a loose fist. He was perfectly motionless. The chair and floor were splashed with liquids of various colors. The man's face was a blank mask, and his eyes were carnelian stones, fixed on his wife.

Ruppol and Raffier inspected the patients but didn't touch them, and they were very mindful of the liquids that were splashed around, since the agent would certainly be in the liquids. The patients were in dreadful trouble, but Ruppol nevertheless felt encouraged by what he saw in the room. This confirmed the reports: The agent caused large amounts of liquids to flow and be expelled from the body. This suggested that the agent was transmitted through contact with blood and body fluids. This relieved Ruppol.

The real worry would be if the agent traveled through the air. If it was airborne, you could catch it by breathing the air near somebody who had it. In which case this hospital room would be very dangerous to anyone who walked into it and inhaled its fetid air. The air of the hospital room would also be dangerous to the human population of the region, because the virus in the room could spread easily to people, who could spread it easily to others who breathed the air near the infected ones.

The patients seemed beyond medical help. Ruppol and Raffier were doctors. It would be extremely unethical to examine a patient and then just walk out of the patient's room with no intention of coming back. They went out of the room and got water and food, and brought it back to the couple. They posted a police guard by the door with orders to let nobody enter the room. This was to prevent anyone else from having contact with the infectious agent inside. Ruppol had no idea *what* was in the room. What was important was that he and Raffier had guessed how the agent moved. It was present in liquids that poured out of the body. To stop it you had to stop people from touching the liquids.

It was getting on toward noon, and they needed to be at the marketplace for the big meeting. When Ruppol, Raffier, and Buassa got to the market, hundreds of people had already gathered, waiting for the doctors to tell them what was going on. The crowd was agitated and confused, gripped by fear. Ruppol borrowed a table from a vegetable seller and stood on top of it so that everybody could see him. He waited a few moments for the crowd to fall quiet, and then he began to speak. He used Lingala, a regional language. *"Today,"* he said, *"we will talk about a disease."*

THE ANCIENT RULE

MARKET, BUMBA VILLE
Noon, October 5, 1976

"*It is a serious, transmissible disease,*" Ruppol continued, speaking in Lingala, as he stood on the table in the Bumba marketplace. "*How does it spread? It spreads through contact with sweat, saliva, and the other liquid humors of the body.*" He had seen these liquid humors in the hospital room, and the humors had come out of every opening in the patients' bodies. "*The disease will be very difficult to stop,*" he went on. "*What can YOU do to stop it? First, you must pay attention to a sick person.*" Then he listed the symptoms of the disease, exactly as he had just observed them in the hospital room. "*You must avoid close contact with a person who has this,*" he said.

"*Now, the second thing you must do,*" he went on, "*is pay attention to the dead. You must not use the traditional method of preparing a dead body for burial. You can gaze on the body of a dead person, no problem there, but you must not embrace the dead. You must not touch a dead body except with rubber gloves, and you should bury the body as soon as possible.*"

Then he went on to recommend a traditional method called the Ancient Rule that people could follow to protect themselves from the unknown disease. As Ruppol wrote in his journal:

By chance, I knew that for many centuries the population of that region had had a customary experience with another disease, smallpox, which was both fatal and highly contagious, and has now been eradicated. Whenever there was an epidemic of smallpox, people who were suspected of having the disease, and their young children, were placed in a hut that was constructed outside the village. The hut was stocked with a supply of water and food, while any physical contact with the victims was forbidden. After a certain amount of time, if anybody in the hut had survived, they were allowed to come back to the village. When there were no more signs of life in the hut, it was burned, with the corpses in it.

"You must apply the Ancient Rule for this new disease," Ruppol told the crowd. He didn't need to explain the Rule to the people of Bumba. They already knew the Rule, just as many of the older people of Bumba knew smallpox. The virus had been eradicated in the central Congo region during the 1960s, though pockets of the virus remained into the 1970s.

Smallpox virus is transmitted from person to person through the air. The particles are embedded in microscopic, invisible drops of moisture that come out of an infected person's mouth when the person is talking or simply breathing. Smallpox is also easily transmitted through contact with pus and scabs.

When smallpox erupts in a person's body, the body becomes covered with tense, pus-filled blisters, called pustules, and a sweet, sickly odor comes off the body. The pain of the pustules is said to be extreme. If the pustules merge into a confluent mass across the person's skin, especially on the face, the person typically dies. If there are no blisters and the skin remains flat but darkens and develops a corrugated or charred appearance, and there is hemorrhaging into the eyeballs or from the orifices of the body, the patient has a 100 percent chance of death. Smallpox is highly contagious. If somebody with smallpox is in one room in a house, people who are in other rooms in the same house can catch smallpox without ever having seen the face of the person with smallpox. If an unvaccinated person catches smallpox, their chance of death is roughly 33 percent, or one chance in three. There is a vaccine for smallpox. The worldwide eradication of smallpox, led by a small team of doctors at the World Health Organization, and accomplished by tens of thousands of vaccinators, is the greatest achievement in the history of medicine.

In 1976, older people still remembered smallpox, and they knew the Ancient Rule. Ruppol assumed that they would teach the younger people about it if the younger people didn't already know. After Ruppol gave his speech in Lingala, he repeated it in French. Then some nurses from the hospital repeated Ruppol's speech in the local language, Budza. *Don't touch anybody who has the disease, don't embrace the dead, bury them right away, and follow the Ancient Rule.*

The next morning Dr. Buassa began going around Bumba Ville educating townspeople about the disease and searching for possible cases. Ruppol and Raffier returned to the hospital to check on the married couple. They also wanted to collect samples of their blood. When they arrived, they found that the guards stationed by the door had left. They looked in the room. Nothing had changed, it was all the same. The wife was still lying on the bed in a fetal position, and her husband was sitting in the lounge chair facing her in exactly the same position as they'd left him the day before, with one foot planted on the floor, his left arm still bent at a right angle, and his fingers curled in a loose fist. They were both dead.

Ruppol and Raffier didn't want to move the corpses far, since they were obviously dangerous. But they had to be buried. The doctors looked around the hospital for somebody who'd help them dig the grave. Nobody wanted to help, so the doctors found a shovel and started digging a sort of trench. But it was a hot day, and they didn't progress far.

The commissioner of Bumba Zone was a man named Citizen Olongo. He worked out of an office in the center of town. Dr. Buassa went to Citizen Olongo's office to see if the commissioner could find them a proper gravedigger. The commissioner sent word into the town that there was an urgent need for a gravedigger at the hospital.

There was no response. The town's gravediggers all seemed to be busy just then.

At that point the commissioner sent word to the town jail: He would liberate any prisoner who was willing to dig the grave at the hospital. To the last man, the prisoners refused. They would rather rot in a Congolese jail than get near a corpse with the deadly illness.

Ruppol and Raffier were at a loss, but then a young man named Mando Lingbanda, who was a janitor at the hospital, came forward and said he'd help with the grave. Mr. Lingbanda got to work, and he dug a deep trench, but as the hole got deeper the doctors began to feel awkward. The janitor, somebody told them, was a person locally described as the "town simpleton." The doctors got concerned and began to wonder if Mr. Lingbanda really understood what he was doing near this virus. I think we should give Mando Lingbanda full credit for what he did. Of course he knew what he was doing, he was digging a grave for some people who had died horribly. Mando Lingbanda goes down in the history of the Ebola wars as the one man in Bumba who had the courage and kindness to give two Ebola victims, a married couple, educators, who had loved and cared for each other and their child to the bitter end, the simple honor of a grave.

After the trench was finished, Ruppol and Raffier stripped off their clothes and put on complete surgical outfits, including disposable paper scrubs, a head covering, rubber gloves, surgical masks, and shoe covers. Then they went into the hospital room to collect the corpses. By this time, the cadaver sitting in the chair had gone rigid with rigor mortis. Given the man's pose, it was obvious that there would be a problem fitting him into the trench. Ruppol tried to straighten out the man's bent arm, but that proved to be impossible. And there was no way he could straighten out the bent leg that had its foot planted on the floor. Finally, very carefully, the two doctors lifted the body off the chair in its frozen position, carried it to the trench, and dropped it in. The body ended up sitting in the grave, with the half-closed fist raised in the air. They placed the woman in the trench, still curled up like a baby. Then Mando Lingbanda shoveled a light covering of earth over the bodies, and the doctors poured gasoline into the grave and tossed in a match, and the grave erupted in fire. The doctors' last sight of the married couple was the man's hand reaching out of the flames as if he were asking for help. The sight haunted Ruppol.

Dr. Buassa remained in the town while Ruppol and Raffier flew in a military helicopter to Yambuku Catholic Mission to check on the situation there. At Yambuku, they found that most of the hospital's nursing staff was now either dead or dying. Ruppol and Raffier toured the Yambuku hospital, and Ruppol headed straight for the dispensary: his usual practice. He suspected there might be a problem in the dispensary. And he was right, he found what he was looking for. It was a metal pan in which several old-fashioned glass injection syringes had been left lying. The syringes had heavy steel needles, a type of needle that can be reused many times. He started asking questions and learned two things. The first was that the nuns didn't know very much about medicine. The second was that they had not been sterilizing the needles of the syringes regularly. The nuns had been giving injections of vitamins and medicines to hundreds of patients each day, and they were using dirty needles. Occasionally the nuns would rinse a syringe in a pan of water to get the blood off it. At day's end, they would sometimes boil the syringes, sometimes not.

Furthermore, Ruppol discovered that the nuns were loaning syringes to hospital staffers who would ride around the countryside on motorbikes, using a single needle and syringe to deliver shots of vitamins and medicines to villagers. They had been visiting villages up to fifty miles away. The staffers on motorbikes had been seeding the virus all over Bumba Zone.

Ruppol, Raffier, and Buassa visited houses of sick people in Yambuku and collected many samples of blood from the patients. They also collected blood from a few people who had survived the illness. It was known as survivor blood. The survivor blood could be used to identify the X virus that was killing people in Bumba Zone. When a person is sick with a virus, their immune system produces antibodies to the virus, which are proteins that stick to the virus particles and kill them.

The antibodies in the survivor blood would *react* to the X virus. This was very important for identifying the X virus.

Ruppol and Raffier ended up crisscrossing Bumba Zone by helicopter, traveling with Dr. Buassa. The doctors visited seventeen towns and villages, and they gave speeches and provided information about the virus in every place they landed, speaking in local languages. *Know what it looks like. Don't touch a person who has this. Don't embrace the dead. Bury the dead quickly. Follow the Ancient Rule.*

They returned to Kinshasa by helicopter, bringing with them a number of vacuum flasks full of glass tubes of blood they'd collected. The doctors transferred some of the blood into new tubes, packed those tubes in dry ice, and shipped them by air to the Pasteur Institute in Paris. The Pasteur Institute quickly sent the blood along to the Centers for Disease Contol in Atlanta, Georgia.

X VIRUS

In midmorning on October 13, Ruppol's blood samples were delivered to the Special Pathogens Branch of the CDC. This is the unit that deals with the eeriest things from the virosphere, the Level 4 demons. In 1976, the head of Special Pathogens was a doctor named Karl Johnson, a tall, bearded man with a soft voice, who had begun his career studying the common cold. Karl Johnson, however, was not a man to be content with mucus. Soon he switched to hunting for unknown jungle viruses that make people bleed. After beating around Central American rain forests, he went to Bolivia, where he discovered a virus that he named Machupo. While Johnson was in Bolivia investigating Machupo, Machupo investigated him, and he ended up in a hospital in Panama where he nearly bled to death.

Johnson had been following the deaths in Zaire with deep interest. He soon managed to get a sample of Sister Myriam's blood. The nun's blood was badly spoiled, just a smear of black ooze. Nevertheless, a scientist in Johnson's team named Patricia Webb put some of this ooze into some flasks of monkey cells, and an X virus grew in the flasks—an unidentified virus. (Patricia Webb was married to Karl Johnson at the time.) In the early afternoon, October 13, an expert in electron microscopes named Fred Murphy photographed particles of the X virus in a tiny droplet of liquid taken from one of the flasks. The particles looked like snakes.

Meanwhile, Ruppol's blood was sitting in a refrigerator at Special Pathogens. It had just arrived that morning. This blood was fresh and red, not black slime. It was perfect for testing—and it included some blood taken from survivors of the disease. This survivor blood had antibodies to the X virus in it. This meant the blood would react to the X virus. But it wouldn't react to other viruses. That afternoon, Patricia Webb, working with a colleague named Jim Lange, discovered that the survivor blood didn't react to any known viruses. Therefore the blood had been infected with an *unknown* virus. And therefore the virus was new to science. Two weeks later the virus would be named Ebola. Patricia A. Webb is credited as the principal discoverer of Ebola virus, along with Karl M. Johnson, Frederick A. Murphy, and James V. Lange.

The CDC team had actually discovered a ticking time bomb. It ticked for thirty-seven years, making very few experts nervous, until it detonated in 2014 and gave the experts a heart attack. Doctors Jean-François Ruppol, Gilbert Raffier, and Buassa played a part by collecting the blood that was the key to the discovery, and by managing to keep it cold using tanks of carbon dioxide gas. The price was the regrettable loss of two crates of French Camembert cheese rotting on the runway at Kinshasa.

Within hours of the discovery, the CDC was preparing to rush an international virus SWAT team to Kinshasa to try to stamp out the virus before it turned into a nightmare for the human species. The director of the CDC appointed Patricia Webb to be the leader of the international team. Soon afterward, though, the director started having doubts about Webb. "This is too big for Patricia," he stated, and he appointed Karl Johnson to lead the team instead. Patricia Webb stayed home in the lab while the men flew off to Zaire. They left Atlanta so quickly that they didn't have time to pack any biohazard safety gear.

A HOLE IN THE NIGHT

KINSHASA
October 18, 1976

A few days later, at the whitewashed house on Avenue Mfumu Lutunu in Kinshasa, Jean-François Ruppol began packing for a second expedition to Bumba Zone, this time with the international team. Karl Johnson, the overall leader of the team, had asked for volunteers and formed them into an away team, called an epidemiology team, or epi team. Their mission would be to travel to ground zero of the disease in Bumba, trace the movement of the virus, and try to break the spreading chains of infection. Ruppol was the epi team's fixer. He knew the country and spoke local languages, and he had already visited Yambuku and seen the disease.

Ruppol packed various items into his travel valise: pipe and tobacco; toothbrush; safety razor; socks. He added a bottle of Johnnie Walker Black Scotch. He thought that the surviving nuns and priests might appreciate it as a stress-reliever. He seemed quite relaxed about the mission.

His wife, Josiane, on the other hand, had lately seemed to him *un peu inquiète*, a little worried. Josiane wasn't quite as relaxed as her husband was about the virus. As she explained to him, what worried her wasn't the virus itself. No, it was the Americans. She was Belgian, and she had a certain view of Americans. A positive view of them *on the whole*. Now it wasn't that they were bad people, it was just that they were *Americans*. And this was a problem. Because, as everyone knew, Americans did not typically know what they were doing. She had begun to imagine that the Americans could do something idiotic with the virus and Jean-François could end up getting infected.

A vision arose in her mind of the Americans returning her husband to her as an infected corpse. But she wouldn't be able to see his corpse (not that she would want to see it if he had died of the virus) because the Americans would have put his body inside some sort of futuristic aluminum coffin resembling a NASA experiment, sealed with rubber gaskets in order to prevent the virus in the coffin from escaping.

She kept her fears to herself, and kissed him and wished him bon voyage, and the children hugged him, and he backed out of the driveway in his car and drove to the military airfield.

Noon, October 19

Hours later, the epi team was in the air, riding in the cargo hold of a Hercules transport aircraft, which was flying at ten thousand feet above the Congo River. The members of the epi team were sitting in jump seats lined up along the windowless walls of the cargo hold. In the center of the hold, in front of them, sat a Land Rover, fueled up and ready to go. Not far from the Land Rover was a bar stocked with whisky, gin, Champagne, and a selection of aperitifs. The plane was a personal aircraft of President Mobutu. The team members were far too keyed up to have a drink.

One of the team members was Pierre Sureau, a doctor from the Pasteur Institute. He had gotten concerned about the fact that the team hadn't brought any biosafety gear with them. While the group was organizing itself in Kinshasa, Sureau went all over the city in a taxi looking to buy some protective equipment, but he soon discovered that Kinshasa was lamentably short of space suits. However, he did manage to buy a dozen French workman's overalls. They were the bright-blue jumpsuits that auto mechanics in France like to wear. He also found some motorcycle goggles. Perhaps if you were a doctor wearing motorcycle goggles and a mechanic's outfit you might impress the locals with the idea that you knew what you were doing around an extreme virus.

The leader of the epi team was a CDC doctor named Joel Breman. During the flight he climbed into the cockpit and chatted with the pilots. It was a hot, windless day. The Congo River, miles wide and divided into multiple channels, had gone mirror smooth and reflected the sky, and stretched ahead of the plane as far as the eye could see, until it was lost in a scrim of clouds on the horizon. Joel Breman thought about what lay ahead. "I was scared out of my wits," Breman recalled recently. "We didn't know what we were facing. We didn't know how it was spread or what we would see in the clinical picture."

Jean-François Ruppol, the team's fixer, had little to say to anyone during the flight. His English wasn't good. He knew how to improve a dispensary at a small hospital in the jungle. He could give people worm medicine and deliver a baby. He spoke Lingala and Kikongo, and he was good at tracking sleeping sickness. But what did he know about Level 4 pathogens that could wipe out millions? He had brought along a rectangular document case made of black leather. Inside the document case Ruppol had placed a secret weapon for fighting diseases. He hadn't told the team what was inside his document case.

The team planned to use the science of epidemiology as their main weapon against the virus. First they would learn the symptoms. Then they would make a case definition of the disease. That is, they would determine how to identify the disease in a patient. Then they would go around finding people with the disease, and they would try to learn how it was transmitted from person to person. Once they knew how it was transmitted, they could start tracing the spread of the virus through the human population. And when they knew who had the virus and who didn't, they would try to block the virus's spread by keeping infected people away from everybody else. In this way, they hoped to eradicate the virus from the human species before it expanded its foothold and became impossible to eradicate or rose into a plague. It was their job to stop the virus before it went medieval.

"Epidemiologists are the steeplejacks of medicine," Joel Breman said. "We're the joiners. We're the people who pull together the steel girders of skyscrapers. Concentration is very important. You're focused on what you have to get done."

The Hercules landed at the Bumba airfield. Ruppol started the Land Rover inside the cargo hold and drove it down a ramp while a crowd of local people shouted, "Oye!"—celebrating the arrival of help. Not long afterward, the team was meeting around a coffee table with the zone commissioner, Citizen Olongo (the man who had earlier offered to liberate any prisoner who would bury the dead schoolteachers for Ruppol and his colleagues). The commissioner said that Bumba had been completely cut off by the quarantine, and it was impossible to get supplies, not even salt or bottled beer.

Ruppol, who was sitting at the coffee table, brought out his black leather document case. He was sitting next to Joel Breman, who wondered what "the Belgian guy" was up to. As Breman later recalled, Ruppol opened the document case and flipped it upside down, and a pile of cash fell on the coffee table—bricks of Zairian notes. It was Ruppol's secret disease-fighting weapon. "Perhaps this will help," Ruppol said coolly to the commissioner of Bumba Zone.

"What the hell? Qu'est-ce que vous faites?"—"What are you doing?" Joel Breman whispered to him. It was a bribe, he thought.

Ruppol shrugged. "It's the way things are done here."

Breman was shocked. "If we start doing this now, we'll never get anything done without payoffs," he said to Ruppol.

This wasn't a bribe, Ruppol explained to Breman. He never offered bribes, never. The money was cash assistance from his organization, Fometro, to an official who needed the money in order to operate his government during the crisis.

October 20

Next morning, the team set out for the Yambuku mission on the dirt road. When they got there, they found the nuns were in a state of deep trauma. Even so, the nuns had prepared a dinner of Flemish beef stew. They were enormously heartened when one of the team members, a Belgian named Peter Piot, spoke Flemish with them and enjoyed their stew. Dr. Ruppol brought out his bottle of Johnnie Walker Black. Father Léo, the priest who made the banana alcohol cocktails, drank half the bottle. After dinner, the team wondered where they would sleep. Any bed at the mission could be hot if an infected person had slept in it. Eventually they decided to sleep on the floor of the girls' schoolroom. They washed the floor with bleach first, then lay down on it.

Ruppol wanted nothing to do with sleeping on any bleached floor. He told the team that he would sleep elsewhere, and he went looking for a bed. After poking around a bit, he ended up at the deserted guest house of the mission, where he found a bed. He drew aside the mosquito net and inspected the sheets. He didn't see any terribly large stains, so he climbed in.

"La nuit fut calme"—"the night was calm," Ruppol's journal tells us. Yet there was something not quite right about the sounds of the night. The jungle was giving off its usual noises, the hoots of colobus monkeys, the low calls of nightjars, the scritching and whining of countless insects . . . all these were the normal sounds of nature in the central African jungle, yet there was something missing from the noise. There was a hole in the night. The sounds of nature were full and complete. What was missing was the sound of drums.

Ruppol had grown up in Zaire. As a boy he had gone to sleep every night with the sounds of drums coming in through the window of his bedroom. Sounds traveled farther after dark, and villagers would use drums at night to communicate over long distances. Often they drummed just to talk, connecting with friends and sharing news, the way people chat on the telephone. Since childhood, Ruppol had liked the sound of drums at night. To him the drums may have been a comforting presence, like your parents talking softly downstairs in the living room while you're falling asleep in your bedroom. But on this night he heard only the sounds of nature, and the sounds of the human presence in the forest had been extinguished. It was as if something so terrible had happened in the forest that people couldn't even talk about it. . . .

. . . Someone was knocking on his door. *"Docteur!* Come quickly please!"

THE ZÁRATE PROCEDURE

Ruppol got out of bed when he heard the knocking. It was still dark, the hour before dawn. He opened the door and found Sister Genoveva standing there. She said that a woman in childbirth had arrived at the hospital. The situation didn't look good.

Ruppol threw on his clothes and got his medical bag, and followed the nun to an open area in front of the hospital, where the woman was lying on the ground on a litter, surrounded by family members. He played a flashlight over her and saw that the whites of her eyes were bright red, suffused with blood. This was a clinical sign of the virus. She was pouring with sweat, gravely sick, and had a very high fever. She was obviously near death.

Ruppol experienced a moment of fear as he looked at the woman. She had a fever, she was at the end of her rope, he thought. There were two lives at stake here. Normally he would do an immediate C-section. Yet a pregnant woman who was infected with the unnamed virus could be extremely infective. If he cut into the mother there would be a lot of blood. And there was the case of Sister Beata, the midwife. She had caught the virus after she'd delivered babies from very sick mothers; mothers close to death.

On the other hand, she might *not* have the virus. If she didn't have the virus, he should not bring her inside the hospital and cut her skin, exposing her bloodstream to the virus. He certainly couldn't do any surgery in the maternity ward, not with blood on the table and the bloody bandages and tampons that were scattered around the birthing room. And the operating room was just as bad.

"The virus is all over the hospital," he said to Sister Genoveva.

He decided to do the surgery outdoors. But he would need an operating table. He asked Sister Genoveva if it would be possible to get a table from the kitchen or dining room and set it on a porch in the open air. While the nun hurried off, he dressed himself in surgical protection—a cotton blouse, a cap, a cloth surgical mask, latex gloves. The virus was transmitted through the liquid humors of the body. He was going to make sure that the woman's body fluids did not have any contact with his skin or with any wet membranes of his body, particularly his eyes and mouth.

Sister Genoveva returned with two men carrying a table. They placed it on a porch, under an electric lightbulb. The men lifted the woman up on the litter and slid her onto the table. She seemed to be in extraordinary pain. One of the men was a nurse named Sukato Mandzoba.

The lightbulb on the ceiling wasn't giving enough light. Ruppol asked the nun and Nurse Sukato to hold flashlights pointed at the woman's birth area. In the beams of the flashlights he could see mucus and what might be small amounts of blood smeared around her vagina. He inserted two fingers into the birth canal and touched the woman's dilated cervix. He could feel the baby's buttocks or feet trapped in the neck of the cervix. The baby was in breech presentation, turned sideways, and it had gotten stuck in the birth canal.

He decided against doing a caesarian section. The procedure would produce large amounts of blood, and the blood could be extremely infective. Furthermore, there was no general anesthetic at the hospital—another reason not to do a C-section. Finally, in Zaire, C-sections were culturally unacceptable. A woman who'd had a baby by C-section was considered to be mutilated, and she could be ostracized by her community. He decided, instead, to do a primitive surgery called the Zárate procedure. Also known as the Zárate cut, the surgery had been developed in the late 1700s in France, and had been modified in the 1920s by an Argentinian surgeon named Enrique Zárate. The Zárate procedure was no longer used in modern hospitals, but Ruppol still did the Zárate cut from time to time, when he thought a C-section would be inappropriate.

He washed the front of the woman's pelvis with iodine and gave her a shot of a local anesthetic directly into the surgical area. She would feel it when he made the incision, for sure she would feel it, but a local would do. He asked Sister Genoveva and Nurse Sukato to take hold of her arms and knees, and grip them tightly. When he made the Zárate cut she could thrash or struggle, and he needed to make the cut very carefully.

The nun took hold of the woman's arms, and Nurse Sukato grasped her knees and raised them until they were bent. Ruppol took up a scalpel in his right hand and placed his left index finger on the front of her pelvis. He moved his fingertip around slightly, feeling the structure of the bones, until he'd located a spot at the front of the pelvis called the symphysis pubis. It's a hard spot, just above the pubic area, where the bones of the hips come together in front. The hip bones form a joint at the spot, but they don't fuse: The hips are held together by a thick piece of cartilage at that spot.

After he had located the spot with his finger, he told Nurse Sukato to start pulling the woman's knees apart. Use gentle pressure on her knees, he told Sukato. Hold them tightly in case she struggles.

Sukato started pulling on the woman's knees, and Ruppol pushed the point of the scalpel vertically downward into the cartilage at the frontal joint of the hips, the symphysis pubis. Then he started making a cut across the cartilage, extending the cut in a vertical direction along the line between the navel and the birth opening. He stroked the blade gently, working it down into the cartilage, while blood welled out of the incision and ran down toward the birth opening. She didn't struggle; she just wanted her baby out. He continued making light strokes with his scalpel across the cartilage. Meanwhile he listened.

Suddenly there was a *crack*, like a rubber band snapping. It was the sound of the cartilage breaking apart and the bones of the pelvis opening up. The moment he heard the sound he stopped cutting into the cartilage and told Sukato to stop pulling her knees apart. He had left a small amount of cartilage in place, which held the pelvic bones loosely together, like a piece of tape. If by accident he had cut entirely through the cartilage, her pelvis would have fallen open.

The baby had loosened. He slid his hand up through the neck of the cervix and got his hand around the back of the baby's head, and pulled the baby out.

The baby came out in a rush of birth fluids and mucus. Tugging on the umbilical cord, he pulled out the placenta. He cut the umbilical cord, and held up the baby and inspected it.

It wasn't breathing.

Ruppol ripped off his surgical mask, bent toward the baby, and covered the baby's mouth and nose with his mouth, and blew a puff of air into the baby's mouth and nose. He gave several light puffs, inflating the baby's lungs a little bit at a time. If he blew too hard he could rupture the lungs.

Sister Genoveva and Nurse Sukato had taken a step backward and were staring at Ruppol. They saw a look of shock come into Ruppol's face. He had suddenly realized what he was doing. But he kept his mouth pinned to the baby's mouth. The baby's chest rose; the lungs were inflated; and Ruppol took the baby away from his face. The baby let out a cry, exhaling Ruppol's breath. It was alive.

Genoveva and Sukato were staring at Ruppol in horror. His mouth, nose, and cheeks were plastered with mucus and birth fluids mixed with blood that had run out of the incision or had come out of the birth canal. He was quite obviously tasting the liquids in his mouth.

"Doctor, do you know what you have done?" the nun whispered.

"Now I know," Ruppol answered.

He seemed frozen in horror. The witnesses saw how his face gleamed with liquids. He kept holding the baby in front of his face and staring at it. He was following standard procedure. After resuscitating a newborn, the physician should watch the baby for three minutes. This is to make sure that the baby continues breathing on its own. If the baby's breathing stops, the physician gives the baby more rescue breaths.

What else could Ruppol do but watch the baby and give it more rescue breaths if needed? It was too late for the doctor to save himself. There was nothing Ruppol could do, nothing at all that would change the choice he had just made. He had done so many childbirths, he had given rescue breaths to so many infants . . . and in a split second he had forgotten himself and had acted out of instinct. Dr. Ruppol knew precisely what he had just done, because these words appear in his journal: "I had just condemned myself to death."

That morning at breakfast, Ruppol was subdued. It seems that he didn't mention to anyone on the team what he had done. Maybe he felt embarrassed. He had given speeches all over Bumba Zone warning people not to touch anybody who had the symptoms, and then he had dived into the virus with his whole face. As for Sister Genoveva and Nurse Sukato, it seems that they maintained a discreet silence about Ruppol's mistake.

The international team broke up into smaller epi teams and began searching through Bumba Zone, looking for people who had symptoms of the disease. They drove in the Land Rovers to sixty-nine villages around Yambuku, asking questions, looking at people, describing the symptoms of the disease to people, gathering information. The roads between the villages were swamped with mud and almost impassable. Some villages seemed indifferent to the disease, not at all worried, while other villages were deeply frightened, and had cut themselves off from the outside world—the villagers had chopped down trees and let them fall across the road in order to stop any vehicles from getting near the village. This was reverse quarantine, in which the village cut off its contact with the outside world in order to protect itself from a spreading disease. In at least two villages, the doctors discovered houses that had been burned down with dead bodies inside them. People had died of the disease inside the houses, and members of the community had then set the houses on fire in order to incinerate the corpses. At least some of the people had followed the harshest measure of the Ancient Rule.

By the end of their investigations, the epi teams had visited villages in an area about sixty miles across, which had a population of 170,000 people. There were very few new cases of the disease. The virus was already almost gone. The teams discovered nine people who had the disease, five of whom died quickly. They found five other people who possibly had the disease and one person whose blood showed they'd been infected but had survived. It was clear to the investigators, after they'd driven around the villages, that the outbreak in Bumba had already nearly ended by the time the international team arrived. Something or somebody had stopped the virus.

If the international team didn't stop the virus, who did? And how?

The evidence is that the people of Bumba Zone stopped the virus themselves. This happened after they knew how to identify the disease and understood how it was transmitted from person to person. Doctors Ruppol, Raffier, and Buassa played a key role in giving people this information during their earlier visit to Bumba.

These things were very, very hard to do. The Ancient Rule went against a normal person's instincts to protect and care for loved ones. The virus was implacable, and people had to become implacable in order to defeat it. They had to restrain themselves from giving care to sick people. They had to cut themselves off from contact with any person who looked like they might possibly have the disease. They had to stop their traditional expressions of grief for loved ones who died. The people of Bumba did these things. They ostracized the families of sick people and offered them no care. Many individuals, grieving for loved ones, seem to have ended their practice of sleeping next to and embracing the dead. In a few cases they burned houses. Above all, they stayed away from Yambuku Catholic Mission Hospital. Any sensible person could see that the disease had been centered on the hospital, and so if you didn't want to catch it you stayed away. Jean-François Ruppol, after urging the people to do the hardest things, telling them to turn cold hearts on the sick and frail, went and did the opposite, and lost his head over a baby.

BLEEDOUT

J. J. Muyembé was the dean of the medical school, and he had a responsibility for the safety of Kinshasa, a city of two million. The case that threatened the city most was that of the twenty-three-year-old nurse at Ngaliema Hospital, Mayinga N'Seka, who had taken care of both Sister Myriam and Sister Edmonda as they died. She had gotten infected, and then, feverish, had gone around Kinshasa and had contact with many people. As Nurse Mayinga lay dying at Ngaliema Hospital, Muyembé and a team of investigators traced her contacts around the city. They followed at least a hundred people, none of whom caught the virus. All the while, Muyembé wondered if he had also caught the virus from the nun just the way Mayinga did, and that he was about to turn into another Nurse Mayinga himself—he, the dean of the medical school. Now *that* would really scare Kinshasa. Meanwhile Mayinga's blood began throwing clots all through her body—the virus had triggered a generalized whole-body stroke. A South African doctor named Margaretha Isaäcson tried to save Mayinga by giving her a blood thinner, to stop the clotting, but the treatment backfired, and Mayinga went into a profuse bleedout from the intestines. She died in shock, with Dr. Isaäcson at her bedside.

Two days before Nurse Mayinga died, some of her blood was collected in tubes, and the tubes were sent to the Special Pathogens Branch at the CDC. There, researchers put drops of Mayinga's blood into flasks, and grew the virus that was in her bloodstream. Today, the virus is called the Mayinga isolate of Zaire ebolavirus, or simply the Mayinga. It's a particular snippet of the changing Ebola swarm that, for a few days, replicated in Nurse Mayinga. The Mayinga is now immortal, and is stored in frozen water in tiny plastic vials that are kept in superfreezers in the Level 4 labs of the CDC. Nurse Mayinga's place of rest seems to have been forgotten. According to one account, her remains were placed inside biohazard bags and given to her family, and her family buried her somewhere along the Congo River not far from Kinshasa, in the village where she was born.

The death of Dr. Jean-François Ruppol, the director of the Belgian medical mission in Zaire, was first reported on October 27, 1976, on the radio in Lower Congo, the province where he had grown up and had practiced medicine for decades. Dr. Ruppol had made the ultimate sacrifice at Yambuku fighting the virus. He had gotten contaminated, and the disease had overwhelmed his body so quickly that there had been no time to evacuate him to the capital, where he might have had a chance of being saved. A large funeral would take place in Kinshasa as soon as the body was returned, and hundreds were expected to attend.

The radio was wrong. At about two o'clock in the afternoon on October 27, Ruppol, unaware of the reports of his death, parked his car in the driveway of his house on Avenue Mfumu Lutunu. He let himself into the house quietly. The children heard him, and they raced to the door, overjoyed to see him. He dropped his travel valise and picked up the children and hugged and kissed them. Then he went into the living room and fell into a chair. He was utterly exhausted, but otherwise he felt fine.

Josiane entered the room moments later. She already knew he was alive, and she was extremely happy about that. Now her husband was sitting in the living room with his shoes on the carpet. Her first thought was to wonder where his shoes had been. My God, she thought, he has hugged the children.

He stood up and offered to kiss her.

She backed away from him. "No kiss," she said abruptly. Then she went over to the liquor cabinet and took out a bottle of Scotch. "Have a whisky first, before you kiss me." Alcohol would disinfect his mouth, of course. She poured a large amount of Scotch into a glass and handed it to him.

He drank the whisky immediately. He badly needed a drink.

She poured him another. "Have two whiskies! Have three! Take the bottle outdoors and drink it all. And go take a bath."

He went outdoors to the patio and proceeded to disinfect himself with about three Scotches. Afterward, in the bathroom, he soaked in the tub, enjoying the pleasant sensation of being alive and at home. After a little while there was a knock on the door, and Josiane poked her head into the bathroom and whisked away his clothing and shoes. She was washing everything, naturally.

Ruppol knew how close to disaster he had come with that baby. After rescuing the baby and getting fluids in his nose, mouth, and probably in his eyes, he had been absolutely terrified, more frightened than he had ever been in his life.

He couldn't *not* think about the married couple he'd examined in the hospital room, the woman curled up on the bed, the man frozen in the chaise lounge and staring at his wife, and the sight of the man's hand reaching out of the flames in the burning grave. All he had done was to try to save a baby. The married couple had caught the virus trying to save their baby, too. Ruppol had spent days giving lectures all over Bumba Zone, telling the people that they must harden their hearts, they must not touch their loved ones or the dead, or they could catch the disease. That they must follow the Ancient Rule. And then *he* had failed to follow the Rule. And why? It was because his human emotions and his doctor's instinct had swept away the Rule, and he had tried to save a child's life.

After Ruppol had stabilized the baby, he had placed it and its mother in a hammock, in a spot at the hospital that seemed free of the virus. The purpose of the hammock was to keep the mother's pelvic bones pressed together while they healed from the Zárate cut. Afterward, he didn't tell the team about his mistake. He kept a perfectly calm professional manner with the CDC doctors and the Belgian virologists, but terror crawled inside him. At the time, the international medical team had had absolutely no idea that Ruppol was in agony.

A mother and her baby who were infected with the virus would both always die: This was known about the virus. He watched the mother and her child like a hawk. If the baby died or the mother's illness got worse, it would mean he was infected.

He watched them for forty-eight hours. After that time, they were still alive and seemed to be doing fairly well. Ruppol's crisis of nerves passed. He decided that she had had a very bad case of malaria. Malaria could produce some of the same symptoms as the X virus: It turns a person's eyes bright red, it produces a high fever, with diarrhea and vomiting, and there can be blood in the urine and blood coming from the intestines. But you can catch malaria only from the bite of a mosquito, you can't catch it from birth fluids, even if you swallow them and get them up your nose. Eventually the mother and her baby were discharged and went home.

Now, as he soaked in the bath, Ruppol felt only a minor sense of satisfaction about the Bumba affair. The nameless agent was in retreat and was nearly gone. Case closed, pretty much. He and the international team hadn't done much to stop the virus, but he'd saved a baby, anyway. At least he hadn't killed himself, which would have been an embarrassment to the Belgian medical mission as well as to his reputation.

The only thing that he and Raffier and Buassa had done was to give people information, in local languages, about how the virus was transmitted and how to protect themselves. They had involved the community. Once people grasped the nature of the virus, they did what was necessary. The virus had turned out to be a piddle. Now he could get back to doing epidemiology on bad pathogens, like sleeping sickness, which could wipe out whole villages and wasn't under control.

Ruppol began to smell smoke.

He got dressed and went outdoors to the patio to have dinner with his family. That was when he noticed the remains of a fire smoldering in the fireplace by the kitchen. This seemed odd, as it was the usual sort of hot, muggy evening in Kinshasa. "What's burning?" he asked Josiane.

His clothes, she informed him. His shoes as well. Actually, she told him, she'd dumped the entire contents of his valise into the fire—she'd turned his valise upside down over the flames and shaken it out. His toothbrush, socks, razor, comb, pipe and tobacco, underwear, everything had gone into the fire, and then, for good measure, she had dropped the valise onto the fire as well.

He thanked her for doing this. In his view, she had done exactly the right thing. Burn anything that's been in contact with the virus. Follow the Ancient Rule. A simple rule practiced for possibly thousands of years in the Congo Basin. In the first outbreak of Ebola virus, the Ancient Rule had prevailed.

Today, Jean-François Ruppol and Josiane Wissocq live in a small house made of stones in a small village in Belgium, where they enjoy the company of their grandchildren. Ruppol is not interested in publicity. Generally he refuses interviews. He says that he doesn't think very often about the first outbreak of Ebola in 1976, though he does keep up with old friends from his days in Zaire. Some of them have died.

I join the Ruppol family for lunch at their dining table. Josiane, a bright woman with a mischievous smile and a sharp sense of humor, has fixed a generous meal. The Ebola problem was a straightforward matter, Ruppol says. He shrugs when he talks about it. It was just a simple outbreak, a matter of giving people correct information about the virus. You involve the local communities in the fight. As we eat a dessert of *clafoutis* and discuss Ebola, Josiane relates her fear that the Americans would return Jean-François as a corpse inside some sort of unbelievable NASA device. She says she told him to drink all of the bottle of whisky. Imagine how difficult it must have been for Jean-

François, she says, after he swallowed all that *jus de mucus*—mucus juice. "I dropped his valise on the fire, too," she says, brushing her hands together in a gesture of good riddance. After lunch we go outdoors to the patio and sit under a wisteria arbor, where Dr. Ruppol lights a pipe and smiles.

Ebola, he points out, was not like sleeping sickness, which killed a lot of people and was virtually impossible to control. Ruppol's point is that after he and his two colleagues gave some speeches to the local people informing them of the virus's mode of transmission—through contact with liquids of the body—the people took care of it themselves. They didn't touch the dead and they buried bodies quickly. The Ancient Rule stopped the virus in a very short time. As for the question of who first discovered Ebola virus, Ruppol didn't think the CDC or Patricia Webb deserved the primary credit. "Who first discovered Ebola?" he says, smiling and holding his pipe. "The people of Zaire discovered Ebola. They discovered it in their bodies."

As for Dr. Jean-Jacques Muyembé-Tamfun, he never broke with Ebola disease. He doesn't understand why he didn't die in the outbreak of 1976. Not the slightest idea why he is alive today. "I think I am lucky," he says. He had multiple massive exposures to infected blood, including when he cut the piece of liver out of the young nurse and drenched his hand with blood that was saturated with Ebola particles. He had ridden for hours in a Land Rover next to Sister Myriam, pressed up against her, and he would never forget the feverish heat coming off her body or the way the rash moved down her arms. The fact that Muyembé is alive today is one of the many enigmas of Ebola virus. Muyembé's survival is as much of a puzzle to his fellow Ebola experts as it is to himself. "I still don't know how Muyembé and Ruppol survived," Peter Piot recently said. Piot was a young member of the 1976 international Ebola epi team in Yambuku, and is now the director of the London School of Hygiene and Tropical Medicine. "I always think that the absence of bad luck is the most important thing in life," Piot said.

Today Muyembé is one of the world's leading experts on Ebola virus. He is a major figure in medicine in the Democratic Republic of the Congo. He has trained approximately 70 percent of the doctors who currently practice medicine there. Ever since 1976, Muyembé has managed outbreak responses to Ebola whenever the virus emerged in his country. He has gone on expeditions to try to identify the virus's natural host. He has also hunted and tracked monkeypox, a blistering virus that jumps out of jungle squirrels into primates, including humans. Monkeypox could someday change and invade dense primate colonies such as Tokyo, London, and Silicon Valley. In 2009, Muyembé and colleagues, investigating a strange, lethal disease that had broken out in a village called Mangala in Lower Congo, discovered a previously unseen rabies-like bat virus that they named Bas-Congo. When Bas-Congo gets into humans it causes a sort of blend of rabies and bleeding, and it is contagious in humans. Nobody has any idea if Bas-Congo rabies could someday blow up in the human species. Bas-Congo virus may remain a hideous curiosity, or it may go far in people. There are almost certainly other Bas-Congo-like rabies viruses circulating in bats, and someday another one of them may get into people. The only certain thing about emerging viruses is that they are deeply unpredictable. Muyembé has also devoted much of his career toward the battle with the emerging virus HIV, and he is one of the world's leaders in the fight against AIDS in Africa. In 2014, Muyembé was awarded the Christophe Mérieux Prize in medicine for his contributions to research and outbreak response.

PART FOUR

CRISIS IN THE
RED ZONE

SITUATION REPORT

JULY 4, 2014
One day after the death of Nurse Lucy May

In the years since 1976, when Ebola first appeared, our understanding of viruses and infectious diseases has deepened. Today, much more is known about viruses—about the structure of virus particles, how viruses get inside cells, what they do inside cells, how viruses jump host species, how they change over time. The classification of viruses has become more detailed and precise. Six species of Ebola have been discovered and given names. It's possible that more Ebolas will be discovered; no one knows.

Genomics, the science of the genetic code, has become extraordinarily powerful, and it is delivering deep insights into how organisms function, including the human organism. Genomics is also revealing new information about the history of life on the planet, and is uncovering secrets of human evolution and the history of human populations. A highly accurate technique of reading genetic code called deep sequencing has proved to be extremely powerful. Deep sequencing can be used to collect and analyze the genetic code of all the organisms in a given sample—deep sequencing gives a panoramic view of all the life forms that exist in a place in the natural world. For example, a small amount of seawater can be deep sequenced: This reveals all of the genetic code in the water—DNA and RNA—from whatever creatures and viruses are in the water. Deep sequencing has revealed that viruses are everywhere.

At the Broad Institute, Pardis Sabeti and her War Room colleagues were doing deep sequencing of the blood of Ebola patients in order to recover the exact letters of code of the kinds of Ebola that had been replicating in each patient's cells. They were trying to get a panoramic view of the Ebola swarm in all its variations as it infected people in the Makona Triangle. Their purpose was to try to learn as much about the West African Ebola as they could, as quickly as possible, so that they could make recommendations to health authorities about control measures, and so that they could see whether the virus was changing significantly.

Pardis Sabeti planned to publish on the Internet all the genome sequences of all the Ebolas they'd found. This was so that other research groups could study the Ebola code and perhaps make discoveries themselves. "There's a lot of siloing that happens during an Ebola outbreak," Sabeti explained. Siloing is when a scientist guards her data, and doesn't release it for other scientists to see or use. Ebola scientists were well known for being secretive, for withholding their discoveries about Ebola until they could publish their discoveries in a prestigious journal and get credit. Sabeti felt that this practice gave an advantage to the virus. "We want to try to break down silos and encourage everybody to work together," she said.

A second shipment of blood samples from Kenema had arrived at Harvard on June 24. This shipment contained eighty-four tubes of sterilized blood serum. The serum had been collected from sixty-six individuals who had tested positive for Ebola. The individuals had lived in the Makona Triangle or its outskirts. Sabeti's team immediately put the samples into the sequencing process.

By July 1, the team had obtained deep sequences of the Ebolas that had been replicating in the blood of a total of seventy-eight sick people. This was a movie clip of Ebola with seventy-eight frames, enough to see a small portion of the swarm as a four-dimensional image of the virus changing through time. Sabeti and her people made a list of all the mutations that had appeared in the Ebolas that had inhabited the seventy-eight people. The list covered just two sheets of paper. They made photocopies of the sheets and passed them around at the Broad Institute.

CAMBRIDGE
July 1

Sabeti and her team spent the day staring at the photocopied sheets of paper. The problem they faced was very simple and very hard. They were looking at the mutations in the Ebola code and trying to understand the meaning of what they were seeing. It was like staring at a cryptic text on an ancient papyrus in which you can read the letters but you don't know what the words mean.

The Ebola code had been shifting as the virus went from one person to the next—a letter changing here, a letter changing there, random errors popping up in the virus's 18,959 letters of code. The errors were propagating; the swarm seemed to be changing. What did the errors mean? Was Ebola evolving in some way?

One of the people who stared at the sheets of paper is a genomic scientist named Daniel Park, who was working in the War Room Team. He dabbed at the sheets with colored highlighters, putting marks on the letters of code that had changed. "Our first question was, What questions should we be asking?" he recalled. "What would be the most helpful for the people dealing with the outbreak? There was a lot of loud talk. We were walking into each other's offices, saying, 'What do you think this means? Is this a transmission chain? Can we piece together a transmission chain?'" Park said. Sabeti and her team began to detect chains of transmission: They could see how a variety of Ebola had gone from one person to another, and then to another.

They tried to identify exactly how the virus was jumping from person to person. Was it really being transmitted only through contact with fluids, or could the virus also be traveling in other ways, maybe through the air? "Being scientists, we were skeptical," Daniel Park said. "We said, 'Is there some *other* way the virus is being transmitted?'"

As the team studied the Ebola code, Pardis Sabeti was constantly in motion, in and out of people's offices and talking loudly in the hallways with shifting knots of her team members. They were worried by what was happening in Africa. They worried that Ebola could experience a big mutation, could really change its character, perhaps becoming more contagious. Sabeti's voice, clear and vibrant, could be heard all over the sixth floor of the Broad Institute. She had sung in her band, Thousand Days, for eight years, and she used her lungs while she analyzed Ebola code, too.

The team became convinced that the outbreak had started in one place, presumably with the little boy in Meliandou, and that the virus had come from an animal reservoir, most likely a bat. It would have been nice to know whether the virus in the Makona Triangle would respond to vaccines or drugs—except that there weren't any vaccines or drugs that were known to work against any kind of Ebola, mutant or not.

By the middle of 2014, two experimental vaccines for Ebola had been partly developed—vaccines that might or might not immunize people against the virus. One vaccine was called the VSV-ZEBOV, and the other was called the IFN-Alpha vaccine. Neither vaccine had ever been tested in a human. In addition, a dozen or so highly experimental drugs were in various stages of development, and most of them had never been inside a human body. It was widely assumed that most of the experimental Ebola drugs would ultimately fail, either because they were unsafe for humans or because they simply weren't effective. Among the anti-Ebola drug candidates was a compound named ZMapp. It had showed some promise when it was tested in guinea pigs, though it had never been tested in a person. The story of ZMapp begins with a man named Larry Zeitlin, and with sperm.

SPITTOON

Larry Zeitlin today is the cofounder and president of a small bio-tech firm called Mapp Biopharmaceutical, Inc., which is head-quartered in a group of rented rooms in a strip office park outside San Diego. Zeitlin is a soft-spoken man in his forties, with dark hair, dark eyes, and a slender frame, who wears jeans and T-shirts, and he is an expert on using antibodies to defeat infectious diseases.

In 1996, Zeitlin was a postdoctoral researcher in a laboratory at Johns Hopkins University in Baltimore, where he was working with a team trying to develop vaginal microbicides that would kill sexually transmitted herpes virus on contact, or would kill human sperm—for use as a spermicide. Spermicides at the time weren't very good. Sperm, it turns out, is very hard to kill. This might be expected, given the fact that sperm has been achieving success in challenging environments for around six hundred million years.

Zeitlin was working on a spermicide that would use antibodies to kill sperm. Antibodies are proteins made by the immune systems of higher animals. The antibodies drift in the blood of an organism, and they stick to and kill invading microbes that get into the blood of the organism. Zeitlin was using human antibodies, and he was trying to see if the antibodies would stick to human sperm cells and kill them.

Zeitlin did little experiments with sperm and antibodies. He would sit on a stool at his lab bench at Johns Hopkins with a microscope and two small plastic film canisters. He would pop open one of the film canisters. It contained a small amount of mucus collected from a woman's cervix. Using a pipette, Zeitlin collected a small amount of the cervical mucus and dropped it on a glass microscope slide. Next he popped open the other film canister. It contained a dollop of warm semen. ("It had been donated that morning by a college student, probably on his way to class," Zeitlin says. "He got paid ten dollars.") Zeitlin put a drop of semen on the mucus. Then he put the slide into the microscope and looked.

He saw sperm cells swimming madly through the mucus. It is actually one of the ancient wonders of nature. The sperm cells wriggled with unending, fierce energy. Nothing seemed able to stop them.

Next, Zeitlin removed the slide from the microscope and put a drop of saline solution onto it. The solution had antibodies dissolved in it. Then, as quickly as he could, just an instant after he put the antibodies on the sperm, he put the slide back in the microscope and looked.

It was too late. The antibodies had already immobilized the sperm.

The sperm cells had gotten clumped together into large shaking balls. They were frozen in place and would eventually die. Antibodies were roughly a thousand times better at killing sperm than any chemical spermicide. Sperm cells can be thought of as invading microbes. The antibodies had stopped the invaders instantly.

In 2000, Larry Zeitlin moved to California to take a job with a biotech startup company in San Diego called Epicyte, which was developing an industrial process to make antibodies for curing diseases and killing sperm. He packed his things in a Volkswagen GTI and set off for the West Coast. One of Zeitlin's bosses at the Hopkins lab was a mucus expert named Kevin Whaley, who also had a job offer from Epicyte.

Zeitlin and Whaley rented a one-bedroom apartment near the beach in Del Mar and went to work at the company.

The company had an industrial method for growing large amounts of antibodies in kernels of genetically modified corn. By the summer of 2003, however, Epicyte was in trouble. One problem was that anti-GMO activists didn't like the idea of GMO corn that could kill sperm. What if a sperm-killing gene escaped from the GMO corn and got into regular corn? What would happen if people ate spermicidal corn?

Zeitlin and Whaley began to smell bankruptcy coming, and they quit the company just before it did go bankrupt. Now jobless, Larry Zeitlin applied for unemployment benefits, and he started collecting a monthly unemployment check of $1,600.

At this point, Zeitlin started thinking about infectious diseases. Could you make a drug from antibodies that would kill a virus that was invading the human body? Could an antibody drug work fast—almost instantly? Just the way antibodies nuke sperm in a matter of seconds?

Zeitlin began thinking about Ebola. Suppose you could cure Ebola with antibodies? Ebola was weirdly like sperm: After a human body got "impregnated" with a few particles of Ebola, about ten days later the impregnated body would give birth to a ghoulish flood of Ebola particles, spurting, gushing, and oozing out of every pore and hole in the body.

In 2003, Zeitlin and Whaley founded Mapp Biopharmaceutical, or Mapp Bio, with a stated purpose of curing infectious diseases, and they were going to start with Ebola. They later wondered if they'd made a mistake when they named their company Mapp Bio, because it sounded like Crap Bio. For venture capital they drew on Zeitlin's $1,600 a month unemployment benefits. Using this money, they rented a three-room suite in an office park. One of the rooms had a garage door, which made the company immediately pre-legendary.

The company's biggest problem was that it was already broke. After paying the rent on the rooms, Zeitlin and Whaley had absolutely no money left over to buy scientific equipment, which is wildly expensive. They needed scientific equipment in order to have any hope of curing Ebola. There was, however, the bankrupt spermicide-corn company Epicyte. It had become a shipwreck, and shipwrecks strew cargo all over the place. When Epicyte filed for bankruptcy, its laboratory equipment, purchased by venture capitalists for large sums of money, was slated to be liquidated at distress prices at a bankruptcy auction. In preparation for the auction, the equipment got put in a locked storage room. Just before the auction took place, some of Epicyte's equipment, in a dazzling, Star Trek–like process, dematerialized from the locked storage room and rematerialized inside the garage of Mapp Bio. Larry Zeitlin figured that the equipment would be sold for next to nothing anyway, and it might do some good advancing the fight against Ebola.

Zeitlin and his partner wanted absolutely nothing to do with venture capitalists. Instead, they started applying for research grants from U.S. government agencies that were involved in biodefense. The general idea was that antibody drugs would be useful in defense against bioweapons. Eventually they got part of a $300,000 grant from the Defense Advanced Research Projects Agency, or DARPA. This agency is famous for investing in blue-sky research—research that has a high chance of failure but could result in a big payoff if it succeeds.

Their plan was to start growing antibodies to Ebola in GMO tobacco plants (rather than in corn plants). They would see if it was possible to grow the antibodies in large enough amounts to make a drug feasible. They started growing tobacco plants under lights in their rented rooms, and they bought two restaurant-grade food grinders. These are machines that chefs use to puree spinach for soup. Zeitlin and Whaley used the food grinders to puree tobacco leaves into a green mush. Then they refined the mush and extracted Ebola antibodies from the mush, a delicate process that made the offices of Mapp Bio smell like a spittoon. They ended up with very small amounts of antibodies dissolved in saline solution. These antibodies might or might not kill Ebola particles; that part of the research hadn't been worked out.

USAMRIID, FORT DETRICK
2000–2013

While the scientists at Mapp Bio were growing antibodies in tobacco leaves, a space-suit researcher at USAMRIID named Gene Garrard Olinger, Jr., was working with a team, trying to discover antibodies that would kill Ebola particles. Olinger had had the same idea that Zeitlin and Whaley had, that antibodies could be a powerful defense against emerging viruses or biological weapons. In 2000, Olinger got a small research grant, and he and his team members began testing 1,700 different antibodies against Ebola in test tubes, at USAMRIID.

The antibodies that Olinger was testing came from mice. Most of the mouse antibodies did nothing to Ebola particles, but five or six of the different antibodies made the particles clump together and die in a test tube.

Even if mouse antibodies could stop Ebola in a test tube, there was a widespread belief among Ebola experts that an antibody drug wouldn't be strong enough to stop Ebola in the human body. The belief was reasonable. In an earlier experiment, an antibody drug had failed to cure monkeys of Ebola. One of Olinger's scientific advisors told him that if he kept on testing antibodies against Ebola he was in danger of dead-ending his career. "People were saying to me, 'You'll never have an impact with this, Ebola is just a horrible disease,'" Olinger says. He didn't heed this advice. He was following a hunch. As always, the medical research was expensive and grueling. But he kept on testing antibodies.

MAGIC SWORD

When a 3-D image of an antibody protein is constructed using a supercomputer, you see a lumpy nugget that typically has the shape of a "Y," like a fork with two prongs. The Y-shaped nugget drifts in the blood. The tips of the fork are like the teeth of a key, and have a truly vast number of different possible 3-D shapes. If the teeth of an antibody match something on the outside of a microbe, the antibody sticks to the microbe, fitting its teeth onto the enemy the way a key fits a lock. Antibodies are very small. If a human cell was the size of a watermelon and an Ebola particle was the size of a piece of spaghetti, then a single antibody protein would look like a speck of finely ground black pepper stuck to the spaghetti.

Human blood is thick with antibodies. A person's blood is about 2 percent antibodies by volume, and they really do thicken the blood. A droplet of blood large enough to cover a person's little fingernail contains about 50,000,000,000,000,000,000 (fifty quintillion) individual antibody proteins. When a human baby is born, its mother produces a special milk called colostrum. Colostrum is a sort of paste made of antibodies. As the baby swallows the colostrum, the mother's antibodies go straight into the baby's bloodstream, and they thicken the baby's blood. The baby's blood has been primed to destroy any invading life form that could hurt the baby.

When a virus starts multiplying in a person, the person's immune system kicks into action and starts making antibodies to the virus. The person's immune system throws all kinds of shapes of antibodies at the invader, antibodies with hundreds of different kinds of key-teeth. Almost all of the shapes fail to work—their teeth don't fit anything on the surface of the invading virus, and so they don't stick to the virus and can't hurt it. However, a few of the antibodies *do* stick to the virus. These are the killer antibodies, the ones with teeth that fit the lock. The immune system goes into overdrive and starts making huge numbers of the killer antibodies, and they fill the bloodstream, sticking to virus particles, covering the particles, and killing them. The antibodies are wet nanobots programmed to destroy anything biological that gets into the human body that doesn't belong there. At USAMRIID, Gene Olinger and his team were making secret Ebola-killing antibodies and testing them in mice. In San Diego, Larry Zeitlin and his team were also making secret antibodies to kill Ebola particles. There was yet another team working on secret antibodies for Ebola, too.

WINNIPEG, CANADA
2009–2014

In fact, a quiet race to find a cure for Ebola had begun. The Public Health Service of Canada has a state-of-the-art Biosafety Level 4 laboratory facility in Winnipeg called the National Microbiology Laboratory. The chief of pathogens of the Canadian lab was then a French Canadian Ebola expert named Gary P. Kobinger. Kobinger, working with a small team, was testing antibodies against Ebola, too. Thus there were three teams trying to cure Ebola with antibodies—led by Larry Zeitlin at Mapp Bio, Gene Olinger at USAMRIID, and Gary Kobinger at the Canadian lab. They knew they were in a race to find a cure for Ebola, and they kept their formulas secret.

Lisa Hensley was then working at USAMRIID, collaborating with Olinger on Ebola research, and she knew him well. She also knew Kobinger well, and she began to think that the rivals should start working together, since it might speed up progress at finding a cure. Hensley began urging them to meet and talk. In the summer of 2012 at a conference in Chicago, they took her advice and went to a bar to get some beers. "This is where I guess we both sized each other up, under the guidance of Lisa Hensley," Olinger says. At the bar in Chicago, and afterward, the two men made a tentative decision to share some of their data. Hensley also urged them to talk with Larry Zeitlin, at Mapp Bio in San Diego.

Soon afterward, Larry Zeitlin and Gene Olinger teamed up for an experiment: They infected mice with Ebola, then gave the mice an antibody that Zeitlin had extracted from tobacco mush in his garage-lab. The drug cured some of the infected mice of Ebola.

In the summer of 2013, the rivals were at a scientific meeting in Maryland, and Hensley invited them out for drinks at a bistro called the Green Bamboo. Hensley, Zeitlin, Olinger, and Kobinger sat outdoors at a patio and started drinking. Gary Kobinger, in addition to doing space-suit research, specialized in digging graves for Ebola victims during outbreaks, and he regaled the others with stories of nearly getting killed by rioting people who thought he was doing horrible experiments on the bodies. Hensley left early in order to put her son, James, to bed, but the others talked on. They agreed to pool their secret antibodies, and would try to create one best Ebola drug, a drug to rule them all. This superdrug would be a combination of the three most powerful antibodies they could find. Olinger called it a cocktail of antibodies. Hensley later began referring to the meeting as the Green Bamboo Armistice. The three leaders were now fully collaborating, and they made plans to start testing antibodies in guinea pigs rather than in mice. This was a step up a ladder of testing that might someday culminate in a human test of a drug that might or might not cure Ebola.

At the same time, Gary Kobinger in the Canadian lab was working on the experimental vaccine for Ebola called VSV-ZEBOV. It had originally been developed in the early 2000s by an Ebola researcher at USAMRIID named Thomas Geisbert, along with colleagues of his. Geisbert's team, which included Lisa Hensley, had tested VSV-ZEBOV vaccine on primates, and it seemed to protect them from Ebola infection. However, the team ran out of money. Agencies that provided money for biomedical research had little or no interest in spending money on a vaccine for Ebola. The virus was considered to be easy to control and no danger to the mass population. Gary Kobinger, however, felt that biodefenses against Ebola should be a high priority. He got money from the Public Health Agency of Canada and began testing the VSV-ZEBOV vaccine in his Level 4 lab in Winnipeg.

WINNIPEG
January 14, 2014

Five months after the Green Bamboo Armistice agreement to develop an antibody cocktail drug for Ebola—and just as Ebola was killing the little boy's family in Meliandou village—Gary Kobinger started testing different Ebola-killing antibodies in guinea pigs in his Winnipeg hot zone. Some of the antibodies had come from Gene Olinger at US-AMRIID and some from his own lab. He tested the antibodies in combinations, seeing which combinations were the most potent at stopping Ebola in a guinea pig. Kobinger, Olinger, and Zeitlin decided to name the drug ZMapp. They made plans to test it in monkeys.

Meanwhile Larry Zeitlin, at Mapp Bio, had worked out a deal with a drug-manufacturing facility called Kentucky BioProcessing, in Owensboro, Kentucky, to start making pharmaceutical grade ZMapp—a purified version of the drug. During the spring of 2014, as the Ebola outbreak sharpened, Kentucky BioProcessing began producing very small quantities of ultra-pure ZMapp—for use in animal experiments, not for people.

The manufacturing process was tricky. By June, the Kentucky factory had produced only a small batch of ultra-pure ZMapp. Of this batch, there were six official extra courses of the drug: excess supply, which wasn't earmarked for lab experiments. Each course of ZMapp consisted of three doses of the drug, in three small plastic bottles or vials of frozen salt water containing dissolved antibodies. These extra courses, numbered one through six, were being stored in a freezer at Kentucky BioProcessing and at a secure facility elsewhere. Courses 1 through 6 had each cost about $100,000 to manufacture.

That wasn't all. There was yet one *more* course of ZMapp. An *undisclosed, unofficial* course. In this book it will be referred to as Course

Zero. Course Zero of ZMapp was sitting in a secure freezer at a location somewhere known to the inventors of ZMapp. Course Zero was a hidden, rainy-day supply of the drug.

Officials at Doctors Without Borders had been hearing about ZMapp, and in April they asked Mapp Bio to send some of the drug to Switzerland in case one their international workers came down with Ebola. Mapp Bio then sent one of the seven courses of ZMapp to the World Health Organization in Geneva. This was Course 1 of ZMapp, earmarked for emergency use by personnel from Doctors Without Borders or from any other international medical group. Courses 2 through 6 remained in the freezers in Kentucky and elsewhere.

WINNIPEG
June 2

A week after the first Ebola patient was admitted to the Kenema Ebola ward, Gary Kobinger began a test of ZMapp in monkeys. He and a colleague named Xangguo Qiu climbed into their space suits and went through the airlocks into a hot zone at the Canadian National Microbiology Laboratory. In the hot zone, twenty-one rhesus monkeys were being housed in cages. Kobinger and his colleague injected each animal with a massive dose of Ebola, enough of the virus to guarantee the animal's death. Then, in the following days, as the animals got sick, the scientists started giving them doses of ZMapp, to see if the drug would prolong the monkeys' lives.

Right after Gary Kobinger began the monkey experiment with ZMapp, he flew to West Africa with a group of Canadians on a volunteer mission to work with Doctors Without Borders. Just then the Doctors were building an Ebola treatment center—an encampment of white tents—in a town in Sierra Leone called Kailahun, which is in the Makona Triangle, about seventy miles from Kenema. Kobinger planned to set up a portable blood testing lab at the Doctors' camp, and after that he planned to work as a gravedigger at the camp.

Kobinger decided to bring some ZMapp with him in case he caught Ebola from a corpse. If this happened, he figured he would make himself the first human experiment with ZMapp. The drug had never been inside a person, and there was no way to tell whether it was safe. In fact, as Kobinger knew, there was a decent chance the drug would throw him into allergic shock and kill him in five minutes. At least it was better than dying of Ebola. On June 18, Kobinger packed the three bottles of frozen, ultra-pure ZMapp into a white foam cooler. This was Course No. 2 of the drug, intended for emergency use by him or any member of his Canadian team.

A few days later, Kobinger and his colleagues arrived at the Ebola Treatment Center of Doctors Without Borders in Kailahun, Sierra Leone. Kobinger placed the foam cooler containing Course No. 2 inside a freezer in a laboratory tent at the camp. After setting up the blood-testing equipment, Kobinger went to work digging graves. By early July he had buried many dead bodies.

The virus was beginning to spread more widely. By July 2, there had been 413 cases reported in Guinea with 303 deaths. In Sierra Leone, there were 239 cases with 99 reported deaths. The virus was now starting to hit hard in Liberia, too, where there had been 107 reported cases with 65 deaths. Ebola had been working its way through Conakry, the capital of Guinea, for a while. And now the virus was beginning to appear in Monrovia, the capital of Liberia.

FREDERICK, MARYLAND
July 1

Lisa Hensley, at the Integrated Research Facility at Fort Detrick, had been watching the growth of Ebola cases with apprehension. By the middle of June it was clear to her that the virus was going to blow up in Liberia. She began to make plans for a second tour of duty with the Department of Defense, to go back to the blood lab at the former chimp station. She had been in constant contact with the Liberian technicians there; they were running out of basic supplies, and they needed more equipment. They were testing more blood every day, and finding more Ebola.

James's school was ending and summer vacation was about to start. She signed him up for various summer camp programs at the local YMCA, and engaged her mother and father, Mike and Karen

Hensley, to come to Frederick and stay with James while she was deployed overseas. As she talked with her son about her plans for West Africa, she thought he didn't seem worried about her this time. In fact, he was annoyed. He asked her how long she was going to be gone *this time*. She made a deal with him: After she got back from Africa, they would take that trip to South Carolina and have a summer vacation on the beach.

Hensley's friend Rafe didn't like the idea of her going back into the outbreak just as it was exploding. He asked her to reconsider her plans to deploy. He was afraid she would take risks and would expose herself to the virus. Rafe had children of his own. He said to her that James should come first for her, and that she shouldn't risk her life trying to save other people when she had James to think about.

But Hensley thought that it was important for a parent to set an example. People needed help, and she was an expert in Ebola. She had seventeen years of experience in space suits handling dangerous pathogens, including smallpox. She would never put herself at risk, never. If she chose not to help and turned her back on people who were dying, what lesson would James take from that? Children miss nothing.

Hensley started collecting space suits and gear. She packed everything in footlockers, and the Department of Defense made flight arrangements for her.

Then, on the first of July, Hensley got an email from her friend and colleague Gary Kobinger, who was working as a gravedigger at the Doctors' camp. The news was stunning.

WOW

I t was evening in West Africa, and Gary Kobinger was sitting with his laptop computer in a plastic tent, just outside the red zone of the camp, and he opened an email from Xangguo Qiu, his colleague in Winnipeg who had been testing ZMapp in monkeys. She had incredible news. She had just finished the monkey experiment. Eighteen dying monkeys had been given superlethal doses of Ebola. And ZMapp had cured all eighteen, rescuing some of them when they had been only hours from death. In some of the dying animals, ZMapp reversed the symptoms of Ebola in ninety minutes.

A ninety-minute cure for Ebola? A hundred percent efficacy rate, even when the animals were hours from death? Results like this simply don't happen in pharmaceutical research. It seemed impossible. Kobinger started emailing people who had contributed to the development of the drug, telling them about the results. They quickly began referring to the monkey experiment as the Wow experiment, because everybody was typing *Wow* in their emails. But the Wow experiment was unpublished. Nobody knew about it except the people who were in the email chains. And the inventors of ZMapp understood that humans aren't monkeys. Just because ZMapp cured monkeys didn't mean the drug would do the same thing in a human. It might do very little. It might kill a human. It might kill Ebola. Nobody knew what ZMapp would do inside the human body.

KENEMA GOVERNMENT HOSPITAL
First week of July

In the days after Wahab the Visioner predicted that many nurses and an important doctor would die at the Kenema hospital, the hospital went into total crisis. Nurses working in the Ebola ward started skipping work. They still hadn't received their $3.50 a day in hazard pay from the government of Sierra Leone. By this time there were three Ebola red zone wards at Kenema Government Hospital; collectively they were known as the Ebola Treatment Center. The three wards held, in total, somewhere between seventy and ninety people infected with Ebola, a number that was steadily climbing, even as many patients died and their bodies were removed. Ebola patients were coming from all over eastern Sierra Leone, riding in taxis or clinging to the backs of family members driving motorbikes. Some of the Ebola victims being carried on the motorbikes were already dead. The bike rider would dump the body at the hospital gates and speed off.

Of the three red zone wards in the Ebola Treatment Center, the first was the original Ebola ward, formerly the Lassa ward, with a normal capacity of twelve beds. This ward now contained about thirty Ebola patients.

The second ward was the Annexe ward, where Lucy May had worked. It was situated near the Ebola ward, and had been converted into a ward for holding patients who were suspected of having Ebola but hadn't yet gotten results from blood tests.

The third ward was a large tentlike structure, with white plastic walls and a metal roof, which had been constructed in late June in a field on the hillside below the Ebola ward. It was called the Tent. By early July the Tent held around thirty Ebola patients lying in cots. The three wards were being tended by just a handful of nurses, who were running out of disposable hazmat suits and equipment. In fact, a severe shortage of hazmat suits had developed at the hospital.

The beds in the Tent were of a certain type that is common in African hospitals, known as a cholera bed. A patient with cholera suffers from uncontrollable watery diarrhea. A cholera bed has a plastic-covered mattress with a hole in the center. A bucket is placed on the floor under the hole and the patient defecates through it into the bucket. This system didn't work well in the Tent. There was a lot of splashing when a patient defecated into a bucket, and patients fell out of bed, got deranged, and wandered, knocking over the buckets. A small number of Ebola nurses were working inside the Tent, emptying the buckets, but it became impossible to keep things clean.

There was a plastic window in the Tent so that people could see and talk to their loved ones inside. Family members of Ebola patients stood behind a protective fence and looked into the window. There were shouts of joy and surprise when a patient came to the window and family members could see the person was alive, and there were cries of sorrow when news came that someone had died in the Tent. Some in the crowd were silent, baffled by the moon suits worn by the nurses and doctors.

Humarr Khan was doing rounds through all three wards.

Michael Gbakie, the epidemiologist and biohazard control officer, was Khan's deputy, the number two operations manager of the Lassa program. As the situation grew chaotic, Michael stayed close to Khan. Khan carried three cellphones. As Khan prepared to enter the red zone, he would hand his phones to Michael and step inside the cargo container staging room, and would put on his suit and gear. There was a full-length mirror in the container. He would stand in front of the mirror and stare at himself, turning himself around, inspecting his gear for flaws. He called the mirror "the Policeman." He said to a reporter from Reuters, "I'm afraid for my life, because, I must say, I cherish my life." After inspecting himself in the mirror, he would round all the Ebola wards.

Khan's three phones rang constantly while he was rounding the red zones. Michael Gbakie would answer the phones and jot messages on slips of paper. When Khan finally came out of the wards and took off his gear, Michael would hand him a stack of paper slips.

People were calling him from everywhere—government officials, doctors, officials from the World Health Organization, international scientists, officials of nonprofit medical aid organizations, families and patients in Kenema. Khan was also making calls everywhere, seeking help for Kenema. Khan and the district medical officer, Mohamed Vandi, spent hour after hour, day after day, on their phones pleading with the minister of health for the promised $3.50 in hazard pay for the Ebola nurses. They made calls all over the government in Freetown looking for the money, which simply didn't arrive. The Ebola nurses were earning their regular five dollars a day, but after the horrifying death of Lucy May, many of the Ebola nurses couldn't bring themselves to go into the wards for that amount.

On July 7, Dan Bausch, Humarr Khan's friend and mentor, who'd recruited him to lead the Lassa program ten years earlier, arrived at the Kenema hospital as a WHO volunteer. Bausch, Khan, and the U.S. Navy doctor David Brett-Major started making rounds together through the Ebola wards. They tried to make sure that dehydrated patients were getting IV saline and that all the patients had access to drinking water and some food, if possible, but that was about all they could do. Every surface inside the Tent and the Ebola ward was covered with a film of Ebola particles.

Khan was under strain. He became irritable, exasperated, prone to flashes of paranoia. He became angry with Lina Moses for reasons that seemed very obscure. He would disappear for hours at a stretch, and nobody seemed to know where he was. "I'm going off to take a nap," he would say to colleagues. He was still seeing some patients in his private clinic. Possibly he felt he could help his private patients, when he couldn't help anybody in the Ebola wards. Who was Khan taking care of? To some observers, it seemed that Khan was failing his nurses.

In early July, a WHO doctor named Timothy O'Dempsey started working in the Kenema Ebola wards. Later he published a description of what he witnessed.

The ETC [Ebola Treatment Center] was poorly maintained and disorganized, with grossly inadequate attention to infection prevention and control (IPC) procedures; evidence of numerous structural and procedural breaches; and catastrophic standards in hygiene, sanitation, and management and disposal of medical waste and corpses. Essential treatment supplies for patients and personal protective equipment (PPE) for staff were woefully inadequate and only haphazardly available. Staff morale was at an all-time low; the nurses were on strike. . . . As they witnessed increasing numbers of their colleagues fall ill and die, [the nurses felt] a growing sense of inevitability that they would be next.

Humarr Khan was the physician in charge of the Ebola wards. Did he look like an incompetent, cowardly doctor, a man not up to the job? Was he failing his nurses and staff, failing the patients? He was getting blamed for the horror in the wards; maybe he deserved it. He was the doctor in charge.

Alex Moigboi, the senior Ebola nurse, continued working twelve-hour shifts in the Ebola ward and in the Tent, going back and forth between the two wards. A sensitive, gentle person, Alex had devoted himself to caring for Nurse Lucy May when she was in the nook at the back of the Ebola ward. Alex and another night nurse named Edwin Konuwa were tending sixty Ebola patients by themselves, all night. There was no electricity in the Tent, and at night the ward went totally dark. Alex continued working with patients in the dark tent at night, talking with them, trying to comfort them, feeling around with his hands in the dark. He seemed unable to leave the Ebola wards. Then Nurse Edwin Konuwa broke with Ebola and ended up in the Ebola ward as a patient. Alex cared for him, too.

On July 9, Alex Moigboi and Humarr Khan put on PPE and went on rounds through all three of the Ebola wards, tending patients, placing IVs. Late in the day, they exited, decontaminated their gear, and removed it. Afterward, Alex confessed to Khan that he wasn't feeling well.

Khan then placed the backs of his hands on Alex's neck—a quick way to check somebody's temperature. Alex's neck felt hot and sweaty: he had a fever. It was probably malaria, Khan thought, and he decided to check Alex's eyes. He pressed his fingertip on one of Alex's lower eyelids and pulled the eyelid downward, which revealed its wet inner membrane. It was bright red. "Look up," he said to Alex, holding the eyelid down. When Alex looked upward, the lower white of his eyeball came into view. The white was red and inflamed, the blood vessels bright and swollen. He checked Alex's other eye and saw the same thing. Alex's eyes revealed classic signs of malaria. Next, Khan took out his wad of currency, peeled off some bills, and stuffed the money into Alex's hand: "Take this money and go buy some malaria medicine, and go home and rest for the night." Alex thanked Khan and headed off to buy the medicine.

CONFESSION

Later, Khan began thinking about what he'd done. He had acted on instinct, the habits of a doctor working in his outpatient office. "Oh," he said to himself, "I have touched Alex." He had gotten Alex's tear fluid on his fingertips. He had gotten Alex's sweat on the backs of his hands. He had pressed money into Alex's damp hand. His hands might have gotten smeared with millions and millions of particles. Had he washed his hands? *When* had he washed his hands? People touch their own faces and eyes all the time, unconsciously. Had he touched his face or eyes before he washed his hands?

The next day, Khan dropped by the office of Simbirie Jalloh, the program coordinator, and sat down on a wooden chair next to her desk. The chair was a kind of worry chair for everybody in the program, where they poured their troubles out to Simbirie.

"Simbirie, I made a mistake," Khan said to her. He explained what he'd done with Alex.

Simbirie felt extremely frightened when she heard him describe the incident. She tried to sound calm. "Don't worry so much, Dr. Khan."

After a few moments, Khan went off to attend other matters.

That same day, Baindu Kamara, one of the three volunteer nurses

who'd assisted Auntie in the failed attempt to save Lucy May, broke
with Ebola disease.

The day after that, Alex told Khan that the malaria medicine wasn't
working. Khan advised him to get a blood test for Ebola. Alex did have
Ebola; they put him in a private room in the Annexe ward.

After Alex broke with the virus, Khan seemed to get ever more
desperate. A tsunami of Ebola cases had started flowing out of Kenema, heading for Freetown, the capital. He dropped by Simbirie's office again and sat down in her worry chair. "Simbirie, I want our
government to quarantine this district."

She thought he looked incredibly tired.

"People are going out of the district and they're bringing Ebola
into the rest of the country," he went on.

"Dr. Khan, I've been calling the government again and again."
The government wasn't listening to her, and the $3.50 in hazard pay
for the nurses still hadn't arrived. "I don't know what I can do," she
said to Khan.

Auntie also dropped by Simbirie Jalloh's office and sat in the
wooden chair. "I am not feeling too well, Simbirie. I am hurting all
over my body."

"Auntie Mbalu, you have to rest," Simbirie said to her. "You're
tired. You've lost your husband. You must give yourself rest."

"What can I do? People are dying."

"Auntie Mbalu! You have to rest!"

Auntie walked back up the hill to the Ebola ward.

July 12

Joseph Fair, the scientist who had originally been planning to work
with Lisa Hensley setting up a blood-testing lab for Humarr Khan,
was now living in Freetown and working as an advisor to Sierra Leone's Ministry of Health. Fair had begun to worry intensely about
Khan, and he decided to visit Kenema and see what he could do to
help.

Fair had first gotten to know Humarr Khan in 2006, when he arrived in Kenema as a graduate student to do research for his dissertation. He got a room at the Catholic Pastoral Center on the outskirts of the city. Soon after he arrived he became sick, and ended up in bed there with a high fever, bleeding from his intestines and throwing up blood. He had also lost the ability to speak except in a faint whisper. A devout Catholic, Fair asked a priest at the center to give him final confession and the last rites. The priest phoned Humarr Khan instead.

Soon afterward, Fair saw a white Mercedes with spinner hubcaps pull up in front of his room, and Khan stepped out. He was wearing his white baseball cap, which gave him a sporty look. He came into the room and examined Fair. "You'll be fine," Khan said jovially. Then he left the room for a moment, and forgot to close the door behind him. Fair heard Khan say to somebody, "This guy is dying! I can't have an expat die on me!"

Khan then returned to Fair's room and started setting up an IV infusion of antibiotics.

"I think I'm passing," Fair whispered to Khan. "I need to make confession. Can I do it with you?"

Fiddling with the infusion line, Khan agreed to take Fair's confession. Fair told Khan about various regretful acts. Khan gave him absolution—he said God would forgive him. Fair also mentioned that he didn't have a will. Khan got him a piece of paper and a pen. Fair scrawled a few lines. He had no worldly possessions, but he left his body to science, and asked that it be autopsied by a space-suited virologist in a Level 4 hot zone, just in case he'd been killed by something interesting. Khan then signed Fair's will as a witness. In a fairly short time, however, Khan's antibiotics began to kick in. Fair recovered nicely, and the two men became close friends and drinking buddies. Fair discovered that Khan was deeply religious, though he was very quiet about it.

Now, after a five-hour drive from Freetown to Kenema, Joseph Fair pulled up by the Ebola ward and started looking for Khan. It was pouring rain, and he couldn't find him anywhere. Eventually, Fair ended up at the plastic viewing window in the Tent, and he saw Khan working inside. By this time the Tent held around forty Ebola patients. "Dr. Khan was working alone in PPE with only one nurse, also in PPE," Fair recalled. "There was blood, feces, vomitus, and urine all over the floor." He didn't get a chance to speak with Khan.

Joseph Fair also went looking for Auntie, with whom he was very close. He found her in the foyer of the Ebola ward, and they embraced. Fair walked around the outside of the small building. Behind the Ebola ward, outside a back door, he discovered twenty dead bodies lying in the drenching rain. They were uncovered, not in body bags, and the rain was carrying body fluids away from the bodies. The people had died in the ward. The back door was open, and Ebola patients inside the ward could see the bodies lying there.

Fair asked Auntie what was going on.

The hospital had run out of body bags, she told him. "We have no access to body bags throughout the country," she said. Fair promised to get her some bags. When he returned to Freetown, he arranged for an air shipment of two hundred body bags from Geneva to Humarr Khan and Mbalu Fonnie.

Monday, July 14

Two days later, in midmorning, Simbirie Jalloh was sitting at her desk in the program office trying to decide what to do with herself. She wondered if she should just leave. People were running away, deserting the hospital. Her mother had been calling her and urging her to leave Kenema and come live with her in Freetown. Her mother was terrified she'd catch Ebola, but she had been saying to Simbirie that if Simbirie got sick with Ebola she must come to Freetown anyway, and her mother would care for her. *Come to Freetown even if you are changed to stone*, her mother had said to her, using a proverb.

Humarr Khan walked into her office and sat in the worry chair. The government of Sierra Leone had just declared him a national hero for leading the fight against Ebola for the nation. Khan didn't seem particularly happy about this, and he said to Simbirie that the virus was going absolutely out of control. A wave of Ebola was coming out of Kenema and was going to hit Freetown, the capital. Again Khan mentioned quarantine.

"Dr. Khan, I don't know what I can do."

Khan had been calling every medical relief organization he could think of, with no results. He also had been thinking about how he had touched Alex Moigboi's eyes. Alex had tested high positive for Ebola, which meant he was probably going to die. "I don't think Alex will make it," Khan said to Simbirie. "If Alex dies, I'm worried for my life." He took hold of his right arm with his left hand. "I would cut off this arm, Simbirie, if I could save Alex's life."

Thinking about Khan's many exposures to the virus, Simbirie started to cry.

Khan got a stern look. "No'r cry, Simbirie," he said in Krio. Don't cry. He switched to English. "When something is really bad, you just do what you can do."

She kept on crying.

He softened. "No'r cry all de time, Simbirie."

Something in his manner frightened her. He was sitting stiffly on the edge of the chair, a few inches away from her desk. No part of his body was touching her desk. He wasn't reaching out his hand to comfort her. He didn't seem to want to touch her or touch anything that was connected to her. Abruptly he stood up and walked out of her office.

Simbirie Jalloh stayed at her desk, in tears. Why had he sat that way in the chair? Why did he not want to touch her desk? Where had he gone? She decided to try to find him.

THE SMOKING PLACE

KENEMA GOVERNMENT HOSPITAL
Late morning, Monday, July 14

Simbirie Jalloh ran up the hill to the Ebola ward, but he wasn't there. She looked into his cargo-container office, and he wasn't there, either. He could be in the general wards, she thought, and so she hurried around the wards, but she couldn't find him anywhere. She was starting to panic. Then it occurred to her that he might have hidden himself, and she ran down the hill along the dirt road to the construction site of the new Lassa ward, and she looked behind the cargo container. There she found Khan, sitting on his plastic chair and drinking a bottle of Sprite.

"Are you all right, Dr. Khan?"

He warned her to stay back from him. "I must keep my distance from you, Simbirie. I don't know if I'm infected or not. We don't know who the next person will be."

She started crying again.

"Now no'r cry all de time, Simbirie." He was holding the soda and looking at the concrete structures and an expanse of lush weeds, green in the rains. "This world is full of mystery," he said. "When you are strong and healthy you do your part to help. And if you fade away, the next man will do his part."

"Let's pray, Dr. Khan."

Khan put down the soda and stood up. Keeping well apart from each other, they raised their hands up in front of their chests, palms turned to the sky, and prayed.

He sat down. She urged him to stop getting near Ebola patients.

"Who else will care for these people? You must do the job and pray." He sipped his Sprite. "You are getting too exhausted."

"I am not exhausted."

He thought she should let her mother take care of her. "Go to you mama na Freetown and go stay fo two weeks."

She refused.

He gave her a slight smile. "Doctor's orders. Two weeks of rest."

Simbirie left Khan sitting on his chair and walked back to her office. The next day she left for Freetown and moved in with her mother. She felt like a coward who had run away from a battle. Even worse, she had deserted her boss, left him alone, sitting on a chair in a corner of the hospital, and he was in a fight that might claim his life.

About the same time, Monday, July 14

The British WHO doctor Tim O'Dempsey had been working at the Kenema hospital, doing whatever he could to help in the Ebola wards. Sometime in the middle of the day he found Auntie lying on top of the table in the nurses' private room, resting, as she sometimes did. This time, though, there was an IV pole next to the table, and Fonnie's brother Mohamed Yillah was watching over an infusion line that ran into Fonnie's arm. He was infusing her with a malaria drug.

O'Dempsey was alarmed. He went out and found Dan Bausch and told him what Auntie was doing. Bausch hurried to the nurses' room. Keeping his distance from her, he gently suggested she should get a blood test for Ebola.

Auntie agreed. Bausch told her he'd do it immediately. He put on gloves and drew blood from her arm. Then he carried the blood tube down to the Hot Lab and handed it through the door to the people inside.

While Bausch was at the Hot Lab, Lina Moses walked into the

nurses' room wanting to talk with Fonnie, and she saw Fonnie lying on the table with the IV in her arm and her brother Yillah watching the infusion. She took Yillah aside. "What's going on with her?"

"She's in for a blood test," Yillah answered.

A wave of dizziness hit Moses. This couldn't be happening. Not Auntie Mbalu. Her next thought was about Auntie's brother. He'd been driving her around on his motorbike. And he had just placed an infusion needle in her arm, exposing himself to her blood. "Are you protecting yourself?" she asked Yillah quietly.

He didn't answer. But she saw a look in his eyes, and it told her everything. He had Ebola, too. He had caught it from his sister.

They put Auntie in the private room in the Annexe ward where Alex Moigboi was dying. He was Auntie's senior Ebola nurse, and now they were dying together. Auntie's brother Yillah stationed himself in their room and began giving them nursing care, doing everything he could to try to help his sister and Alex get through Ebola alive. He didn't bother to wear any sort of biohazard protection when he cared for them. He didn't need any protection from Ebola now.

CALLER I.D.

Lisa Hensley worked a twelve-hour day in the Liberian National Reference Lab at the former chimpanzee station. She stayed in her space suit almost all of the time, working in the negative-pressure hot zone, testing samples of blood using the PCR machine that could detect the genetic code of Ebola. In late morning she exited from the hot zone and stepped into a tub of bleach water, took off her boots, then used a pump sprayer to spray her suit with bleach, and took off her suit. Underneath she wore blue cotton surgical scrubs, surgical gloves, and socks. She stepped into her loafers, then went into a room that had a balcony that looked over the metal roofs of the chimpanzee enclosures. It was raining, and the rain was hammering on the metal chimp houses, making a lot of noise.

She ate some peanut butter crackers and drank some bottled water. That was lunch. She made some calls on her government phone, and then suited up and went back into the lab, and worked until she couldn't stand up anymore. She was working with an Army colleague named Randal Schoepp. There were times when she or Schoepp had to exit from the hot lab due to gastrointestinal issues, a common thing in Africa. Every time she felt a little sick, Hensley wondered, faintly, if this was an amplification of Ebola.

The virus was hitting Monrovia hard. All of the city's hospitals were filling up with Ebola patients, and the medical system in the city had become almost nonexistent. A blood sample came into the lab that had been taken from a pregnant woman who was in a distressed childbirth. She was lying on the sidewalk outside the main city hospital of Monrovia, bleeding from her birth canal. The doctors wouldn't admit her because she appeared to have Ebola, and a pregnant woman with Ebola who is hemorrhaging is an extreme danger to medical staff. Hensley knew that the woman and her baby needed immediate medical attention. She tested the woman's blood sample immediately—but the process took two hours. The result came up: The woman didn't have Ebola. She could be admitted; she and her baby could be saved. But by the time she got the results back to the hospital, the woman and her baby had died on the sidewalk.

At night, a U.S. Embassy car took Hensley and Schoepp to their hotel on the beach. The streets felt unsafe; civil order was beginning to break down. At the hotel, Hensley would eat something and then, just before she went to bed, she would call home on Skype. Her parents were taking care of James, and he would be eating supper. She would chat with him over supper, then talk with her parents. Her father, Mike Hensley, had begun to worry about her. He didn't say anything to her about his worry, not a word, but he was a research scientist who had worked to develop drugs for HIV. He had been watching reports of the outbreak and talking with Lisa at least once a day. From her reports he could see that things were going to get much worse in Monrovia. This Ebola outbreak had just begun, in his opinion, and Lisa was in the center of the storm. He knew his daughter very well. He thought she was likely to disregard her safety as she tried to help people. He said nothing about his fears to Lisa or to Lisa's mother, Karen.

KENEMA HOSPITAL
6 a.m., Saturday, July 19

Auntie declined rapidly. Eventually she was moved out of the private room in the Annexe ward and into the Ebola ward. She ended up lying in the cot in the nook at the back of the ward where Nurse Lucy May had died. Nurse Alex remained by himself in the private room.

At dawn on Saturday morning, a nurse entered Alex's room and found that he had died during the night, alone in the room. There was nobody available to put the body in a biohazard bag, so it was left lying on the bed.

At about nine that morning, Simbirie Jalloh, who was living at her mother's house in Freetown, called Humarr Khan. He didn't answer. This was unusual. He had caller I.D., and he always answered her when she called.

She waited an hour and called him again. Still no answer. She began calling all three of his cellphones, in case one of them had run out of charge or something. There was still no answer. She kept calling his phones all day, getting more and more worried. Finally, after dark, she phoned the ambulance driver who drove Khan to the hospital in the mornings. "What is happening with Dr. Khan? I have called him ten times."

Doctor had stayed at home today, the man said.

She realized that he had been seeing her number coming up on his caller I.D. all day, but he hadn't been answering. Why? Why wouldn't he answer? A terrible fear gripped her, and her body began shaking. Then her phone rang. It was Khan. "I have a fever," he said.

It was a brief call. Afterward, Simbirie Jalloh couldn't stop shaking. She told her mother that she would return to Kenema immediately. She could not desert Khan after all.

SCREAMS

Next morning, while Simbirie Jalloh was riding in a taxi on her way to Kenema, a blood technician visited Humarr Khan on Sombo Street and drew a sample of his blood and brought it back to the Hot Lab. Just as the Hot Lab people had started testing Khan's blood, Auntie went into cardiac arrest in the Ebola ward. An Ebola nurse named Alice Kovoma, who was tending her at the time, shouted for help, and she and the WHO doctor David Brett-Major started chest compressions on her and managed to resuscitate her. Auntie continued breathing on her own afterward, but she was in a coma. Alice Kovoma and another Ebola nurse named Nancy Yoko stayed at Auntie's bedside, doing whatever they could to keep her alive. They couldn't rouse her to consciousness.

The news that Auntie Mbalu's life was hanging on a thread in the Ebola ward raced through the city of Kenema. The city revered her. Within minutes, worried people began coming in through the hospital gates and gathering in front of the Ebola ward, anxious for any news about Auntie's condition. Many nurses and hospital staff were mixed in with the crowd, wearing civilian clothes—they had stopped working at the hospital. The crowd grew larger, and it began to get agitated.

People felt that the nurses' candlelight vigil had failed. Wahab the Visioner's prophecy had come to pass. More nurses were dying of Ebola, and now Kenema's Auntie could die. The crowd got bigger, and louder, and shouts came out of the crowd. Auntie could not die. Incompetent fools on the hospital staff had let the virus spread all over the hospital, and now it had gotten into Auntie. If Auntie died, there was going to be payback, they shouted.

At about one o'clock a rumor went through the crowd that Auntie wasn't in a coma, but that she had been dead for many hours. The rumor had it that the hospital staff was hiding the truth, afraid to announce the news for fear the hospital would be attacked by the mob.

There were young men in the crowd, hotheads. They began shouting that if Auntie was dead and the staff was lying they would burn the hospital to the ground. The nurses had brought the virus to Kenema, the men shouted. The hospital staff was responsible for all the deaths in Kenema. And now they had killed Auntie. The young men whipped the crowd into a mutinous rage, and it grew larger and larger as more and more people joined the mob. The crowd had the numbers and the passion to destroy Kenema Government Hospital.

In the crisis center in the Library, Lina Moses heard the shouting and noise of the crowd around the Ebola ward. Mbalu Fonnie and her late husband Richard Fonnie had a teenage daughter named Martiko. Moses suddenly realized that Martiko was probably near the Ebola ward waiting for news of her mother. Martiko could be in danger, she thought, because the crowd was sounding violent. Lina ran out of the crisis center and up the hill to the Ebola ward, shouting, "Martiko! Martiko! Where are you?"

She found Auntie's daughter alone in the crowd, with nobody to protect her, and she was sobbing. People in the mob were shouting that Auntie was dead.

Lina wrapped her arms around the girl and held her. "Mbalu isn't dead," she said to the girl. "If your mother was dead I would know."

Just then, a series of piercing shrieks came out of the Ebola ward. Two nurses were screaming.

Lina knew immediately what it was. She kept her arms wrapped around Auntie's daughter.

The crowd paused in its agitation, and grew momentarily quiet, listening to the screams coming out of the ward. The nurses' screams went on and on, an aria of unfathomable loss. As the cries fell over the crowd, their meaning began to sink in. Auntie had just died. The nurses' cries wouldn't stop, and they seemed to cool the crowd like falling rain. Sounds of weeping started in the crowd, gusts of sobs, until the entire crowd was in tears, and the mob collapsed in upon itself, all of its anger gone. The shrieks of the nurses kept coming out of the ward.

Lina Moses began to worry that something dreadful was happening inside the building. She told Auntie's daughter that she'd be right back. And then she ran straight through the doors of the Ebola ward and into the red zone.

There was no time to put on safety gear, not even rubber gloves. Lina ran down the narrow corridor, past the cubicles, headed for the screams, trying not to notice too much around her, trying to keep herself focused on getting to the nurses. The smell was beyond description. Patients were naked, or were lying in soiled clothing. She saw red eyes, people near death, people dead, filth and liquids on the floor, which was slippery under her hiking boots. Moses had always told herself that she had good instincts for knowing where the virus was and where it wasn't, and now she knew where it was. The virus was everywhere around her, on every surface, every bed, every person, every wall. The ward was a melted reactor core of virus, and she ran through it toward the screams, trying not to touch anything.

LAST OFFICES

EBOLA WARD
1:30 p.m., July 20

When Lina Moses reached the nurses they had moved away from the nook where Auntie lay, and they were crying out uncontrollably. Moses knew them well—Alice Kovoma and Nancy Yoko. The two nurses seemed paralyzed with shock and grief. They couldn't look at the body of Auntie or go anywhere near it.

Moses took the nurses lightly by the arms, touching the sleeves of their Tyvek suits. She said that their cries were upsetting the crowd, and she persuaded them to move inside a room. Then she closed the door of the room and went back through the ward and out the door. She had been inside the ward for just a few minutes, but she had touched the nurses with her bare hands. Their suits were heavily contaminated, because they had been working on Auntie, trying to save her.

As soon as she got outdoors, Moses went to a barrel of bleach water that was standing outside the Ebola ward, and she washed her hands in the water. Then she returned to Auntie's daughter, who was sobbing. Lina Moses didn't know what to do, she felt useless around death. It occurred to her that maybe the girl would like a soda. There was a man who sold soft drinks, and she thought he might be nearby. "We need a Fanta!" she shouted.

The soft drink man heard her, and he brought two cold bottles of Fanta and opened them. Moses led Auntie's daughter down the hill to the unfinished buildings of the new Lassa ward, and they sat down not far from Khan's smoking chair and drank the sodas.

Michael Gbakie watched them as they walked, and he saw how they were crying. He felt like crying himself, but he couldn't. He was Humarr Khan's deputy. He had been working at the hospital, getting two or three hours of sleep at home, then riding his motorbike to the hospital in the middle of the night, dealing with constant emergencies. Khan was at home getting tested for Ebola and Auntie was dead. *When everybody is in tears,* he thought, *somebody has to see to what is happening. What are the things that need to be done?* Someone had to do the last offices for Auntie. Feeling as if he was the last man standing on a battlefield, Michael donned a set of PPE and entered the ward. He found Alice Kovoma and Nancy Yoko standing in the room where Moses had left them, crying and terrified. Now that Auntie was dead, nobody was safe. We're going to die, they said to him.

He agreed with them: They were all at risk. "You know, it is not easy," he said to the nurses. "It is not easy." They would have to prepare the body because they were the ones who were left alive and were able to do it.

Alice Kovoma finally agreed to go with Michael to the body. When she said she'd help, Nancy Yoko said she'd help, too. The two nurses and Michael went into the nook where Auntie lay on the cot. They sprayed her body with bleach and moved it into a body bag, leaving the bag outside the Ebola ward to be collected later by the morgue team.

There is something more to be said about the two nurses who tried to save Auntie. Alice Kovoma was a slender, beautiful woman in her forties, with a sparkling personality. She was well-liked and admired, and she had worked in the ward for many years. She had tried to save Dr. Conteh as he died in the ward, and had wept over him. Five days after she tried to save Auntie, she tested positive for Ebola. Two weeks later, on August 5, she died in the Ebola ward among the patients, in the place where she had served.

Nurse Nancy Yoko continued to work in the Ebola wards. Later in the outbreak, she became the director of Ebola nursing at the Kenema hospital: She replaced Auntie. One day during the crisis, Yoko told a British colleague that she intended to work in the Kenema wards until the virus was extinguished. "I will not go far from Ebola," she said to her British friend. "I'm here until we get Ebola finished in this country. I have the faith, that is all." Not long afterward, on September 14, 2014, Nancy Yoko developed symptoms of Ebola disease, and she died a week later in an Ebola ward among her patients, just as Mbalu Fonnie and Alice Kovoma had done.

At about nine o'clock on the morning after Auntie died, a Hot Lab technician phoned Simbirie Jalloh. "Dr. Khan is positive," he said.

An explosive anger came over her. She felt that the entire hospital and Dr. Khan had been abandoned by international authorities. Abandoned to die.

Shortly afterward, Khan phoned the Ministry of Health and informed them that he had Ebola. Half an hour later, one of his phones rang. It was the president of Sierra Leone, Ernest Bai Koroma, wanting to speak with him.

FREEZER

President Koroma told Humarr Khan that he had just spoken with Dr. Margaret Chan, the director-general of the World Health Organization, and that she had supported a direct request for help from the WHO in getting Khan evacuated from Sierra Leone. If Khan could be flown to Europe or the United States, he would receive the best possible medical care. He also might have access to experimental drugs and treatments that weren't available in Sierra Leone.

The minister of health, Miatta Kargbo, was on the line. "We will have a plane waiting on the tarmac for you," she said to Khan.

Joseph Fair, Khan's friend, who was working at the Ministry of Health, was also on the line, listening, though he didn't speak. As the call ended, Fair privately asked the government officials if he could have a word with Dr. Khan. The president and the minister of health signed off. Fair waited for a moment and then identified himself. "Cee-baby, it's me, Joseph. How are you feeling?"

"I'm feeling tired. I'm not acutely sick yet."

"I need you to stay strong and remember how much we've been through together." Fair knew that Khan had important decisions to make, so he got ready to say goodbye. But Khan wanted to talk about something.

He told Fair that he'd been reading up on experimental treatments for Ebola. He had been looking for some drug or vaccine that might help his patients. And now he was the patient. He had concluded that there were three plausible options for treatment. There was the experimental vaccine called VSV-ZEBOV (this was the vaccine that Gary Kobinger and his colleagues were developing in Winnipeg). Khan understood that the vaccine might save a person's life even if the person was already infected with the virus. In addition, there were two drugs that showed some promise, in Khan's view. One was called TKM-Ebola and the other was called ZMapp. "If it was up to me, I would comfortably take any of these," Khan said to Fair. He added that he was leaning toward ZMapp. The drug had been able to cure some guinea pigs. He asked Fair what he thought.

Fair didn't know what to say. He had *heard* of ZMapp, but he didn't know anything about it. He didn't even know how it was supposed to be administered to a human. "I have no idea what it would do to you," he said to his friend. "You might go into anaphylactic shock and die in five minutes. And then you would be Experiment Number One."

"What would you do, Joseph?"

Fair silently interrogated himself. What would he do if he had Ebola? "It's a miserable fucking death," Fair explained later. "I've seen it when it was my friends dying. Humarr had seen his friends die of Ebola, and he knew what was coming. There's a saying among Ebola experts that if we get Ebola we'll fall on a needle. Any needle."

After some hesitation, Fair gave Khan an answer. "I personally would take ZMapp."

Khan wanted to know more about the drug. "Will you send me the scientific papers on ZMapp?"

Fair promised he'd gather up the scientific papers on ZMapp and send them to Khan in an email. At that moment, what neither Joseph Fair nor Humarr knew about were the eighteen monkeys that ZMapp had cured no matter how sick the animals had been. The Wow experiment had been finished just three weeks earlier and hadn't yet been published. Only a few experts knew about it. Humarr Khan also didn't know about Course No. 1 of ZMapp, which had been deposited in a freezer in Geneva.

10:45 a. m., July 22

After the phone call with the president, Khan made a decision to be transported to the Ebola Treatment Center of Doctors Without Borders in Kailahun, Sierra Leone, a town in the Makona Triangle. He didn't want to be placed in a ward in Kenema where conditions were horrific and where he would be cared for by his own nurses, who would be demoralized to see him sick. The Doctors' camp was equipped with elements of basic care. It also had European and American managers and doctors, who were working alongside paid local staff, who were mainly Sierra Leonians. Khan figured he could spend a short time in the Doctors' camp until he could be airlifted to a hospital in Europe or the United States.

His three phones were ringing constantly. He was losing energy, getting sicker. He didn't want to talk on the phone. He gave some money to his houseman Peter Kaima and told him to go to the shops and buy him a cheap cellphone. It would have a new number, a secret number. He would share the number only with a handful of people. He especially didn't want his parents to find out that he had Ebola. His mom and dad were in good health, still living in their small house on the edge of Freetown Bay. His dad was ninety-nine, though, and Khan feared the news could kill him.

He could hear a crowd gathering in the courtyard. He stayed indoors. An ambulance arrived, and the crew, wearing PPE and carrying a stretcher, entered the house. They found Khan lying on his bed, fully dressed. They helped him put a bio-hazmat suit over his clothing, and a mask and gloves, so that he wouldn't spread the virus to others. Then they carried him out of the house on the stretcher. When the stretcher men emerged from the house carrying Khan dressed in a hazmat suit, the crowd erupted with cries and sobs.

Khan asked the ambulance men to stop, and he got off the stretcher and stood up. "Gentlemen, don't worry. I'm going to come back," he said to the crowd. Then he walked to the ambulance and climbed into the back and lay down on the gurney. Khan's houseman got into the ambulance with the driver. It pulled out of the courtyard and went north along Hangha Road. An hour later, driving on pavement, the ambulance crossed the Makona River on a bridge. The pavement ended after the bridge. The ambulance slowed down and began creeping along a dirt road that wandered into the Makona Triangle.

MONROVIA, LIBERIA
11:35 a.m., July 22

While Khan's ambulance was leaving Kenema, Lisa Hensley was traveling through Monrovia in a four-wheel-drive vehicle owned by the U.S. Embassy, with an Embassy driver. She was on her way to the Eternal Love Winning Africa Hospital, situated along a beach on the Atlantic Ocean just south of the city. Hensley and her Liberian colleagues had been testing blood samples that were being sent to them by doctors at ELWA Hospital, and Hensley wanted to visit the operation. An Ebola ward, being run by the Christian medical relief organization Samaritan's Purse, had been set up in the hospital's chapel.

The Embassy vehicle stopped near the chapel, and Hensley got out. The chapel, a small, whitewashed building with a cross on it and a wide-spreading roof, was surrounded by white plastic barriers, and it was full of Ebola patients. Hensley met the head of the Ebola ward, a physician with Samaritan's Purse named Kent Brantly—a thin, tall American in his thirties, with sandy hair and a neatly trimmed beard. Samaritan's Purse medical staff, Liberians and Americans dressed in PPE, were going in and out of the chapel through designated entry and exit points. At the ward's exit point, the medical workers were having their suits decontaminated with bleach spray by decontamination staff.

Kent Brantly was very busy. He needed to go inside a supply store-room to work for a while. Hensley offered to help. They went into the room together and moved boxes of medical supplies from place to place as they talked. Afterward, they went out into the sunshine, and Hensley took a picture of Kent Brantly with her phone. That evening, at her hotel in Monrovia, she took out her phone and looked at the pictures she'd taken that day. Kent Brantly had seemed fine in the storeroom, but as she looked at the photo she'd taken of him, she noticed that he had dark circles under his eyes, and he looked terrible. It had been dark in the storeroom when they were talking and working together, and she hadn't noticed how tired he looked.

POTTER'S FIELD

An hour after Lisa Hensley took a snapshot of Kent Brantly, Pardis Sabeti got an email from Robert Garry, the Tulane microbiologist who had been gathering blood samples in Kenema for the Ebola genome study. Garry reported that Humarr Khan had Ebola. After reading the email, Sabeti walked into the Ebola War Room for a scheduled meeting. "Within thirty seconds," Sabeti later recalled, "my face began to melt, and I just began hysterically crying, and then I looked up to see that everyone else was crying, too."

She called Robert Garry after the meeting. He was in his office in New Orleans. Sabeti and Garry had been making arrangements to set up an international conference call to discuss delivering experimental drugs and vaccines to African healthcare workers. Nurses were dying in Kenema—Mbalu Fonnie was dead, too—and Sabeti and Garry had been getting frantic as they tried to find some way to protect them from the virus. They decided that the first topic of the conference call would be to get medical help for Humarr Khan himself.

KENEMA GOVERNMENT HOSPITAL
Simultaneously—1 p.m.
While Sabeti and Garry were talking on the phone, an ambulance pulled up in front of the morgue at the Kenema hospital. Two men wearing PPE emerged from the morgue carrying the body of Auntie Mbalu Fonnie, which was inside a white bag. They placed her in a casket, and loaded the casket into the ambulance. A few minutes later the ambulance arrived at the Kenema cemetery, which is situated in a brushy area on the outskirts of the city. It serves as the city's potter's field and was the only place where the Ebola dead were permitted to be buried.

Many graves at the cemetery are unmarked. Grave markers, where they exist, are typically a wooden board or two pieces of wood fashioned into a cross. Mbalu Fonnie had always expected to lie next to her husband at the foundation of their house on the lower slopes of the Kambui Hills, a house he had built for Mbalu and himself. That would not be possible.

The Ebola graves had been dug with haste. The bodies had been buried in shallow holes, with a mound of dirt on top. It had been raining hard, and the rains had washed out the mounds, revealing white body bags that poked out of the soil. Dogs or rats had torn open the bags and dragged out body parts. Ribs and arm bones were scattered in the weeds, and a human femur poked out of one of the mounds. Auntie's mother, Kadie, was devastated by the sight. The family stood well back from Auntie's grave as men wearing biohazard gear lowered her casket into a shallow hole, and covered it with earth.

As he was riding home in a taxi after the funeral, Fonnie's brother Mohamed Yillah began feeling warm. He knew this was his Ebola coming on, so he asked the taxi driver to stop at the hospital for a few minutes so he could have his blood drawn for an Ebola test. Afterward, at the family compound, Yillah told the children of the household not to touch him. He shut himself in his bedroom and asked family members to leave food for him outside the door.

· · ·

Later that afternoon, Humarr Khan's ambulance arrived at the Ebola Treatment Center of Doctors Without Borders, in Kailahun, in the Kissi Teng Chiefdom of Sierra Leone. The treatment center was a cluster of tents sitting in a tract of forest on the outskirts of town. It was jammed with patients, and more were arriving every day. The staff at the camp was deeply stressed, hardly sleeping, and starting to get overwhelmed. When Khan arrived at the camp he was treated just like everyone else. Someone drew a sample of blood from his arm, and he was assigned to a tent in the red zone of the camp.

An Ebola treatment center of Doctors Without Borders is the Ancient Rule updated with plastic. A hut made of palm fronds, placed apart from the village and stocked with water and food, was an ancient red zone. Smallpox victims were isolated in the red zone hut and were kept from having any contact with people in the village. There was no treatment for smallpox or for Ebola. You went into the red zone, you got food and water, and you either survived or you didn't. After the survivors emerged from the hut, the hut was burned. This was on-site disposal of biohazardous corpses, the same thing the Doctors did when they buried infected corpses next to the camp.

At a typical Ebola Treatment Center of Doctors Without Borders there were a dozen isolation tents made of white plastic, lined up in the center of the red zone. Each tent had cots for twenty patients. A plastic basin was placed next to each cot, which the patient could use for vomiting. There was a line of pit toilets—privies—near the tents. A disposable pad was placed on the cot if the patient had diarrhea and couldn't move. The pad was supposed to be changed by a worker once a day. The patients were given plates of food and bottled water and soda. Bright orange plastic fences surrounded the red zone. There was a visitors' area, where friends and family members could speak with patients across the fence. The visitors were required to stand at least six feet back from the fence. An Ebola patient could vomit suddenly, throwing infective droplets six feet through the air.

Khan was placed in a sort of semi-private tent, which had only six cots in it, which was reserved for local medical workers who got infected. The part of the red zone that was next to the visitors' area was covered with a metal shade roof that was supported by wooden posts. Khan sat on a plastic chair under the shade roof next to the visitors' area, made some calls to government officials on his secret cellphone. He was very sparing in the use of his phone. There was no electricity in the red zone, and he had no way to recharge his phone when the battery ran down. His phone couldn't be taken out of the red zone to be charged, since it was contaminated with Ebola particles.

The sun went down, and Khan prayed. He didn't feel very sick. He had only a mild fever and body aches. He still had a pretty good appetite. No vomiting or diarrhea. He had seen this before in many patients. The disease can start out like a minor bug, and you hardly know you have it.

A few lights went on at the camp. A generator was humming somewhere, supplying power to two laboratory tents, which were stocked with blood-testing machines and freezers. A hundred feet away from Khan's tent, inside a freezer, there was a cube made of white Styrofoam. The cube was about eighteen inches on a side, and it was dented, battered, and grimy, sealed with packing tape. Nobody at the camp knew what was inside the foam cube or who had left it in the freezer. The staff of the camp were much too busy to pay any attention to the cube. Inside were three small plastic bottles of frozen water with antibody proteins dissolved in it. They were Course No. 2 of ZMapp, and had cost $100,000 to manufacture.

After he had finished his work at the camp, Gary Kobinger had left the drug in the freezer and returned to his lab in Canada. He had left the drug at the Kailahun camp because he wanted to test the stability of ZMapp in a tropical climate. He had no intention of offering the drug to anybody. ZMapp had never been inside a human body.

KAILAHUN ETC
Morning, Wednesday, July 23

"Nobody's telling me anything," Khan complained to Simbirie Jalloh the next morning, speaking with her on his secret phone after his first night at the camp. He was wondering about plans for air evacuation. He told her that he had forgotten to bring his passport with him. She told him she'd take care of it. Khan spent much of the morning sitting in a plastic chair in the visitors' area of the red zone, listening to rain clatter on a metal roof overhead.

Simbirie called Michael Gbakie and asked him to search Khan's house and bring his passport to him. "Please stay with Dr. Khan. He needs a friend with him," she said. Michael found the passport in Khan's bureau, and, hours later, he arrived at the visitors' area of the camp. He told Khan he'd keep the passport nearby until Khan needed it—the passport couldn't be brought into the red zone because it would get contaminated. Michael got a room in the town and stayed.

As afternoon arrived in Sierra Leone, it was morning in the United States. Pardis Sabeti, in her office at Harvard, and Robert Garry, in his office at Tulane, prepared to start an international teleconference call to discuss getting experimental drugs and vaccines for African health-care workers. Khan's need for an experimental drug was going to be their first order of business when the call started. Sabeti and Garry had looked into the dozen or so untested drugs that might be used to treat Khan. They had come to the conclusion that ZMapp would be the best option for him—if he could get flown to someplace where he could be offered the drug.

Pardis Sabeti had been planning to lead off the call, but as the moment approached, she got afraid that her voice might break as she talked about Khan's situation, so she asked Garry to speak first. Sabeti and Garry waited a few moments as Ebola experts in many places around the world began identifying themselves and joining the call. All of them knew Humarr Khan personally, and several of them were his close personal friends. The meeting went live.

TELECONFERENCE

9:30 a.m., Eastern Daylight Time, July 23

Robert Garry began by saying that if Humarr Khan could be evacuated to a hospital in Europe, he should be offered experimental drugs. In Garry's view, ZMapp was the best option. The drug had been successful in monkeys but had never been tested in a person. Khan was a medical doctor who had done trials of experimental drugs on human patients. Therefore he could weigh the risks of taking a drug like ZMapp, and could make an informed choice. "If there's anybody who should get this drug, it should be Dr. Khan," Garry said.

Next, Pardis Sabeti spoke. She said that Khan was the perfect candidate for an experimental trial with ZMapp because of his understanding of Ebola, and because he was a national leader in Sierra Leone who could serve to inspire his country. But it was important to not just help Khan. "There has to be a sense of justice in how we proceed. It will be important to keep moving forward on treatments for everybody," she said.

The molecular biologist Erica Ollmann Saphire spoke. She was at home in San Diego, California, in a quiet corner of her apartment, while her husband organized breakfast for their children. Saphire was then running a laboratory at the Scripps Research Institute in La Jolla, where she was studying the molecular structure of Ebola and other viruses. She had collaborated with Khan on research into an anti-Lassa drug. She spoke forcefully, saying that ZMapp would be the best choice for Khan. She had been extremely impressed that the drug had saved eighteen monkeys with no failures, an extraordinary result by any measure.

Khan's friend Dan Bausch spoke. He was in Geneva, having just arrived there after working in the Kenema Ebola wards with Khan. He mentioned that there was a course of ZMapp sitting in a freezer in Geneva. The drug was earmarked for international health workers, but Bausch thought it would be appropriate to remove the drug from the freezer and fly it to the Doctors' camp so that Khan could take it if he chose to. The drug had to be sent to Khan quickly, because he was getting sicker by the hour.

Then Dr. Armand Sprecher, a leader of Doctors Without Borders in Brussels, dropped a bomb. "We can relax and not worry about the logistics of transport," he said. "It's not necessary." Sprecher had recently learned there was *another* course of ZMapp sitting in a freezer right at Khan's camp. Khan didn't need to be flown anywhere.

The news startled many of the people on the call. They had had no idea that a course of ZMapp was already at the camp. Pardis Sabeti, Robert Garry, Dan Bausch, and Erica Saphire felt a sense of relief when they heard that there was ZMapp at the camp with Khan. Garry got the impression that the drug would be given to Khan within hours. They felt the problem with Khan had been settled—he would get ZMapp at the camp. Nobody knew if it would help him, but it might give him a better chance of survival. Once the problem with Khan seemed to be solved, the meeting immediately moved on to the larger problem of getting experimental medicines delivered to all the African medical workers who were at risk from Ebola.

KAILAHUN ETC
2:30 p.m. local time

Minutes after the conference call ended, Armand Sprecher phoned the clinical operations manager of the Kailahun camp, a registered nurse named Anja Wolz. Sprecher told Wolz about the conference call. He said that the experts had recommended that Khan be offered the ZMapp in the camp's freezer. He told her he was personally in favor of it.

Anja Wolz told Sprecher that she wasn't comfortable about offering the drug to Khan. She had just learned about the drug in the freezer herself, and her reaction had been, *"Oh, shit."*

She told Sprecher that the idea of giving a totally untested experimental drug to just one patient in the camp made her feel very uneasy.

"I'm not giving you advice," Sprecher said to Wolz. "Dr. Khan is a friend of mine, but I know you will do the best thing."

She told Sprecher that she would consider the idea of giving ZMapp to Khan. "If you tell me this is okay, I'll think about it," she said to him.

Right afterward, Anja Wolz called a meeting of the camp's doctors and managers. Brussels, she said, was recommending that Humarr Khan be told about the drug in the freezer and be offered it. Among the doctors in the tent was a physician named Michel Van Herp. Van Herp and another doctor raised objections.

The problem for them was fairness. Humarr Khan was a doctor. He was privileged. How was it fair to offer an experimental drug to a privileged doctor, when many other patients—children, poor people—would be dying of Ebola right next to the doctor while he was getting a treatment of extraordinary rarity that might save his life? All the other patients had no chance of getting an experimental drug.

Doctors Without Borders adheres to an unyielding ethical principle known as "distributive justice." The principle asserts that all human beings deserve equal access to the best available medical care. Under the principle of distributive justice, every person is entitled to the same care, whether they are rich or poor, powerful or weak. The principle requires that medical resources must be spread out equitably among all patients according to their needs. A homeless drug addict is entitled to the same medical care as a powerful government minister. In a disaster, if there is a shortage of doctors and medicines, the shortages must be spread out equitably among all patients—distributive justice does not play favorites in a disaster. For many of the European and American workers at the camp, if Khan was given ZMapp it would be an injustice and a breach of their most important ethical principle.

Furthermore, the camp managers were worried that the drug could send Khan into allergic shock and kill him instantly. It had never been tested on a human. There was no oxygen at the camp—no way to sustain his breathing if he shocked out. As Joseph Fair put it later, "They basically said, 'We're not an ICU, we're a fucking tent, and you're asking us to stick ZMapp in him.'"

If Khan died after getting the drug, Africans might believe the drug had killed him. Even if the drug did nothing and he died of Ebola, it might seem to Africans that the drug had killed him. There had been violence near the Doctors' camp, and the camp's managers feared that the local population could attack the camp if an African doctor died there after being given an experimental drug by white foreigners. There was a long history of violence by local people directed against Doctors Without Borders during Ebola outbreaks. Recently there had been rioting and threats of violence around a Doctors' treatment center in Macenta, Guinea, not far from the Kailahun camp. If Khan died after being given the drug, Anja Wolz and the others felt that his death could endanger themselves and their patients, which could put their entire mission at risk.

And what if the opposite happened, what if the drug *saved* Khan? This too would be a breach of ethics, because a privileged doctor would have been saved when no one else could be saved. There were only two possible outcomes of giving the drug to Khan, he would either die or survive. If he died it would endanger the mission. If he survived, and the drug appeared to have helped him, this would be a serious breach of the principles of Doctors Without Borders. Several people at the camp said that if ZMapp was given to Khan it would be so unethical that they would consider resigning from the mission—they would quit and start working somewhere else, or would go home.

They debated whether they should tell Khan about the drug in the freezer. If he knew about the drug he might ask for it. If they refused to give it to him and he died, then Africans would say that the white people had a special drug for themselves but they had refused to give it to a prominent African doctor and so he had died, and there could be violence directed at the camp. It seemed like the safest and most ethical action would be to keep Khan in the dark.

Khan had no knowledge of this roiling debate.

While the camp officials were discussing whether to offer the drug to Khan, Tim O'Dempsey, the WHO doctor who had been working in the Kenema wards with Khan, arrived at the camp in order to check up on Khan and offer him clinical care if he needed it. O'Dempsey was the director of humanitarian programs at the Liverpool School of Tropical Medicine, and is an internationally known expert in the delivery of medical care to populations in crisis. When he arrived at the camp he found that its leaders had closed themselves in a tent and were conducting a private meeting. After an hour, they invited him in and explained the situation.

O'Dempsey knew next to nothing about ZMapp, but he knew Gary Kobinger, a co-inventor of the drug. O'Dempsey had a good cellphone, and he called Kobinger in Winnipeg, and reached him. He asked Kobinger if he himself would take ZMapp if he was in Khan's position.

"Absolutely," Kobinger answered. In fact, he said, he had planned to test the drug on himself if he had caught Ebola.

Anja Wolz spoke with Kobinger. "Gary, I don't know what to do," she said.

Wolz recalls that Kobinger said to her, "You are in an impossible situation. Any decision you make will be the wrong decision. But we stand behind you and the decision of the team."

After talking with Gary Kobinger, Tim O'Dempsey presented his own view to Anja Wolz and the camp doctors and managers. He argued that Khan should be told about the drug's presence in the freezer and should be offered a chance to take it. Khan's risk of dying from Ebola far outweighed any risk the drug posed to him, O'Dempsey argued.

At this point, though, none of the camp's medical staff, including the doctors present, would agree to administer the drug to Khan. It was a unanimous decision: Even if Khan asked for the drug, they wouldn't give it to him.

Tim O'Dempsey had come to the camp with plans to give Khan medical help if he needed it. He asked Anja Wolz if he could put on PPE and enter the red zone so that he could meet with Khan and assist him. She told him that he could not.

As it turned out, Wolz had been one of O'Dempsey's students at the Liverpool School. He got into a long, tough discussion with her and other camp managers as he tried to persuade them to let him go in and help Khan.

I made multiple requests to Tim O'Dempsey for an interview, but he declined them. However, O'Dempsey's colleague at the Liverpool School, Dr. Tom Fletcher (the WHO doctor who had worked as an advance operative at the Kenema hospital) did have a comment. "Tim O'Dempsey had to negotiate hard before he was allowed to see Khan," Fletcher said. "From the reports we got from Tim O'Dempsey, we felt Khan's care could have been better. Most people I know think his care was far from optimum. It upset a lot of WHO staff."

A Canadian doctor named Rob Fowler, who worked with Tim O'Dempsey at the Kenema hospital, said he could understand Anja Wolz's position as well as O'Dempsey's. The conflict between them was a matter of life or death. "Imagine you are a physician," Fowler said, "and you walk up to Dr. Khan's bed and say to him, 'Dr. Khan, we need to have a discussion about your treatment options,' and then you do a three-sixty and you see fifty other people in exactly the same position as Khan. How is that fair? Of course, I'm sure that Tim wouldn't stop at giving treatment just to Dr. Khan. They would have had a hard time getting Tim out of there once they let him in." Fowler said that he'd worked with O'Dempsey in the Kenema wards, and it had been really hard to persuade him to pull out of duty, even when he was exhausted and it was late at night.

Finally, the camp officials allowed O'Dempsey to put on PPE and go into the red zone to see Khan. But O'Dempsey was told that he must not tell Khan about the drug in the freezer.

A doctor has a duty to inform a patient of the treatment options that are available to the patient. If nobody at the camp would agree to administer ZMapp to Khan, then the drug wasn't available as a treatment for him, and so the camp doctors had no ethical duty to tell him about it.

When international medical personnel went to work for Doctors Without Borders, they were informed that if any of them came down with Ebola, the person would be flown to Geneva, placed in a world-class hospital, and offered experimental drug treatment. The treatment, specifically, was human-grade Course No. 1 of ZMapp.

After O'Dempsey got permission to go into the red zone, he suited up and went in, and found Khan upbeat and smiling. Khan said he had a headache and body pains, but he still had an appetite. O'Dempsey asked if there was anything he could do for him. Khan asked for jelly coconut water.

ELWA HOSPITAL, MONROVIA, LIBERIA
Late morning, Wednesday, July 23

While the experts were starting to discuss treatment for Khan, Dr. Kent Brantly, the chief physician of the Samaritan's Purse Ebola ward in the chapel at ELWA Hospital, made an exit from the ward. He passed through a decon line where Samaritan's Purse volunteers, who were wearing bioprotective gear, sprayed him with bleach and helped him remove his gear. Once Brantly had gotten de-suited, he collected his phone and called the director of emergency medical operations of Samaritan's Purse, Dr. Lance Plyler. Plyler was Brantly's boss. "Lance, don't freak out, but I think I've got a fever," Brantly said to him.

Lance Plyler told Brantly to go home and isolate himself. Brantly lived in a whitewashed bungalow on the grounds of ELWA Hospital. He went to bed in his house. A nurse from Samaritan's Purse visited him, drew a sample of his blood, and wrote a fake name on the blood tube, "Tamba Snell." Plyler didn't want anybody to know that the head Ebola doctor was getting a blood test. It was now about twenty-four hours since Lisa Hensley had been shifting boxes with Brantly inside a closed storeroom and talking with him, their faces inches apart.

NIGHTFALL

Hours later, as night fell over West Africa, Humarr Khan lay down in his cot for his second night in the camp. Unknown to him, the international conference call about him was still going on. Doctors with the World Health Organization were getting angry. They were strongly in favor of telling Khan about the ZMapp and giving him a chance to decide for himself. The camp officials opposed this idea and were seeking alternatives. Some managers at the Brussels office of Doctors Without Borders favored offering the drug to Khan, but they felt that the camp managers had authority. Anja Wolz didn't want the camp's staff and patients to be endangered by violence if Khan died. She was skeptical of the drug and afraid it would fail or kill Khan. The discussions were heated and impassioned.

Gary Kobinger, in Winnipeg, stayed on the calls and answered questions about ZMapp. Since he was one of the inventors of the drug, he couldn't make a recommendation—it's illegal for the developer of an unlicensed drug to advocate giving it to a human subject. "I tried to keep my voice neutral," Kobinger recalled. He described how the drug had saved eighteen monkeys from death. Each animal had been given three doses of ZMapp, spaced a few days apart. He compared this to three punches from a prizefighter: The first two punches knocked Ebola down and the third ended the fight.

As the hours passed, WHO doctor Tim O'Dempsey agreed to administer the drug to Khan himself. If something went wrong, the WHO would get the blame from the Africans, not Doctors Without Borders. The plan became unworkable when camp doctors asserted that if O'Dempsey gave ZMapp to Khan they would not be willing to be involved in any aspect of Khan's subsequent medical care. O'Dempsey couldn't care for Khan by himself without any backup.

They came up with a new plan for Khan. At 10:45 p.m., West Africa time, Gary Kobinger, in Canada, sent an email to Larry Zeitlin, the president of Mapp Bio, who was at his office in San Diego. The two men got into a rapid-fire exchange.

> KOBINGER: Ok there was no treatment today/night. WHO HQ is pissed but MSF [Doctors Without Borders] decided against. Just got more lab data, he could still be saved . . . Btw WHO is putting him on medevac, if he can last the ride.
> ZEITLIN: Wow—they sending him to Geneva?
> KOBINGER: Geneva, london, or france, whom ever accepts him.
> ZEITLIN: So they aren't planning to treat him with the Geneva dose?
> KOBINGER: Sending him with 1 dose, rest will need to come from Geneva (or all 3).

The plan was this: A medevac jet would pick up Khan and fly him to some country in Europe. Tim O'Dempsey would travel on the jet with Khan. O'Dempsey would bring along one of the bottles of Course No. 2, the ZMapp at the Kailahun camp. He would administer the bottle to Khan during the flight. If the drug sent Khan into shock, there would be medical equipment on board the plane that might save him. If Khan survived and got to an advanced hospital in Europe, two bottles of Course No. 1—the Geneva course—would be flown to Khan.

That way, Khan would get the required three doses of ZMapp, and his treatment would occur outside Africa.

LIBERIAN NATIONAL REFERENCE LAB
8:30 a.m., Thursday, July 24

An Embassy vehicle picked up Lisa Hensley and her American colleagues at their hotel and took them on the hour-long drive to the national lab. By the time they arrived, a delivery of blood samples had come from Samaritan's Purse at ELWA Hospital, an insulated box containing tubes of blood packed on ice. Each tube was packaged inside a plastic bag, and the bag had been sterilized. Hensley suited up and went into the hot area, and the team began testing the blood samples. One of the samples was from a patient named Tamba Snell. The name was common, and nobody thought anything about it. By the end of the day, the team had found that Tamba Snell was negative for Ebola. Hensley reported the blood results back to Samaritan's Purse. At Samaritan's Purse, when Lance Plyler learned that Brantly's blood test was negative, he ordered a second blood test. Kent Brantly was still isolated at home, and he wasn't getting better.

KAILAHUN ETC
Later that day

The discussions about getting a jet for Khan continued the next morning. Officials at the WHO headquarters in Geneva eventually hired an air medical company called International SOS to fly Khan to Europe. But France, Germany, and Switzerland wouldn't immediately agree to allow a foreigner infected with Ebola to cross their frontiers. The SOS jet would need to land somewhere and refuel, perhaps in Mali or Morocco. No country would allow the jet to land on their soil if it was carrying a confirmed Ebola patient.

Tim O'Dempsey then proposed a different idea to Anja Wolz and the camp managers: He would take Khan back to Kenema himself. He would bring the camp's supply of ZMapp with him. At the Kenema hospital he would personally administer the drug to Khan. If Khan was given ZMapp in his own hospital, Africans would be less likely to think he was the victim of a white experiment. This way, there would be no danger to the Doctors' camp, and the Doctors wouldn't have to breach their ethical code by giving the drug to Khan.

The camp leaders agreed to O'Dempsey's proposal, but they stressed that he must *not* tell Khan about the ZMapp in the camp's freezer. O'Dempsey went to the visitors' area and talked with Khan across the red zone fence. He asked Khan if he would like to return to Kenema. Speaking vaguely, O'Dempsey said that it might be possible to offer Khan treatment in Kenema that wasn't available at the Doctors' camp. He didn't say anything about ZMapp.

Khan didn't want his staff to see him infected with Ebola. He told O'Dempsey that he had more privacy at the camp, and he would stay at Kailahun.

After that, Michel Van Herp approached Anja Wolz privately. He couldn't stand to see what was happening with Khan. "I am going into the center to give the medication to Khan myself," he said to Wolz.

Wolz recalls that she advised him not to do it. "I was like, 'No, from an ethical position you shouldn't do it.' It was the most horrible situation I had ever been in with Médecins Sans Frontières."

Same day

Michael Gbakie, who was keeping Khan's passport for him, visited Khan regularly, speaking with him across the fence at the visitors' area. Khan told Michael that he had come down with diarrhea and was getting dehydrated. He asked Michael for IV hydration—a drip of saline solution in his arm, to replenish his fluids.

The camp staff, feeling overwhelmed by the number of patients, and concerned about the dangers of a bloody needle, had stopped giving IV fluids to patients. Michael told Khan that he would come into the red zone and set up a saline drip for him. He thought the best thing for Khan would be Ringer's solution, which has potassium in it. If Khan's potassium got too low he could have a heart attack. Michael went in search of a camp official who would allow him to enter the red zone with an IV hydration kit and administer it to Khan.

He found a doctor and asked to speak with him for a moment. He told the doctor that Dr. Khan was getting dehydrated and needed IV hydration. He wanted to go in and set up a drip of Ringer's for Khan.

The doctor answered that nobody was allowed to go in except staff of Doctors Without Borders.

Michael identified himself as Dr. Khan's deputy, and said Dr. Khan needed assistance. He said he was very experienced with PPE.

Just then, the doctor happened to be with the camp's logistician. The logistician, who is one of the most important officials at the camp, manages the supply chain and physical operation of the camp. The situation at the camp was dire, and the logistician seemed exasperated with Michael's request. "Why is everybody focusing on Dr. Khan?" he said. And no, Dr. Khan could not have any extra treatment that wasn't available to the other patients. The other patients were not getting IVs. Dr. Khan could not have an IV if the other patients weren't getting them.

It seemed crazy to Michael. "Do you actually know what you are saying?" he said to the logistician. "If every patient has to be treated equally, okay. But do you know how many lives he has saved? And how many lives he will save if he is alive?" If an infusion of saline solution could save Khan, it would enable more lives to be saved.

Michael claims that after he asked the question, the two managers turned their backs on him and walked away without a word. He would not be allowed to go into the red zone with an IV hydration kit for Khan.

JUSTICE

KENEMA GOVERNMENT HOSPITAL
Six months later

I am sitting in Michael Gbakie's office, not far from the Hot Lab. It's a hot day in January, at the height of the dry season. A wind called the Harmattan has filled the air with dust blowing in from the Sahara Desert, and the sky is the color of a lion's fur. The virus is still flickering in Kenema, but the great fire has died down. Sierra Leone is getting a thousand new Ebola cases a month, a number that is dropping rapidly. The virus is active in Kono, an area to the north of Kenema. The schools of Sierra Leone are closed. All over the country there are roadblocks, where soldiers and police officers point a digital thermometer at your forehead and ask you questions about where you've been and where you're going. A Red Cross Ebola center has been set up not too far from Kenema, and the number of Ebola patients in the Red Cross center has been going steadily down. The Ebola wards at the Kenema hospital have been closed. There are no Ebola patients at the hospital. The general wards are full of patients, and the food vendors move quietly along the walkways.

Michael Gbakie has survived the passage of the virus. He is a quiet man, not tall, with a rectilinear face and an air of sensitive reserve. Through the window of his office I can see the Tent. It is empty.

"Do you feel the camp managers treated you with contempt when they turned their backs on you?" I ask.

He speaks calmly. "Of course, with contempt."

"How did you feel? I mean, what were your emotions just then?"

His gaze drifts sideways, as if he's looking away from something, not at something. "According to how I felt—at that time, emotionally, I had the impression on my face that I was not satisfied with his answer to my question. I didn't shout. I was just a little bit calm. And they walked away. They said nothing, as if I was not somebody."

"Were they white?"

"They were both white."

"Do you feel this was racism?"

His answer surprises me. "No," he says firmly and immediately. "It doesn't sound like racism is the problem here."

I am struck by this. "What is the problem, then?"

The problem, for him, wasn't simple racism. The camp managers seemed insular to him, caught up in a rigid rule and rigid procedures that were interfering with the saving of lives. He was a medical professional engaged in the same fight they were engaged in. He had ten years of experience wearing PPE and managing care for hemorrhagic patients infected with a Biosafety Level 4 virus. He was making a point about justice in medicine, and he didn't like being ignored when he raised questions about their concept of justice.

"There's been a longstanding problem between MSF and the local African healthcare community," said John S. Schieffelin, a pediatrician at the Tulane University School of Medicine who had worked in the Kenema hospital during the crisis. "When I went up to Kailahun to talk with the European staff at MSF, I was treated one way and the African staff we partner with in Kenema were treated very differently. In Kenema, our African staff are treated as equals. The Europeans at MSF didn't engage with our Kenema staff as colleagues. It's wrong. It's just wrong. They're also trying to impose their version of justice on people who see it differently. We couldn't even convince MSF of the use of IV hydration," Schieffelin said.

At the time, the leaders of Doctors Without Borders believed that giving Ebola patients IV saline would not improve a patient's chance of survival. They also felt it would expose workers to bloody needles—an unacceptable risk. When the number of patients surged in the treatment units of Doctors Without Borders, they halted or greatly reduced the practice.

"I find this difficult," Tom Fletcher said, referring to the Doctors' decision to stop giving IV fluids to Ebola patients, "because they didn't administer much IV anyway. Anyone who has used IV fluids on Ebola patients cannot understand the rationale for not using it. At Kenema, we had two or three doctors for up to a hundred patients, and we gave IVs to everybody who needed one. Placing an IV doesn't take that long and doesn't present a high risk to a health worker. Various needles have a cap that flips over to make it safe. Getting people to sip oral rehydration by mouth is much more time-consuming because you have to sit there with the patient. It's also more dangerous, because it's close contact with the patient over a longer time, and the patient can vomit. The solution is not to ban IV use—then you get a case-fatality rate of seventy percent. With a willingness to use IVs, you can drop the fatality rate below fifty percent."

Tom Fletcher worked for a time at Donka Hospital in Conakry, Guinea. The Donka hospital staff gave IV hydration to all Ebola patients who were dehydrated. One day a physician from Doctors Without Borders, who'd been working at the Doctors' Ebola Treatment Center in Guéckédou, visited Donka Hospital. "She started weeping as she looked around," Fletcher recalled. "She said to me, 'How can the disparity of care exist between where I work [at the Doctors' unit] and here in this hospital? We can't utilize our skills at IV because of a policy.'"

When a doctor is working in a disaster, doesn't the doctor have a duty to help as many patients as she can, even when she can't help everybody? This is the practice of triage—of deciding which patients to treat first when you can't treat all of them immediately. Physicians commonly practice triage in a disaster. They work on as many patients as they can, even when they must leave some patients without care.

"Of course it is justice to give IVs to some people and not to others," Tom Fletcher went on. "It's crazy to say nobody gets an IV just because some people can't get it. That is madness."

BRUSSELS
Summer 2015

Dr. Bertrand Draguez, the clinical director of the operational center of Doctors Without Borders in Brussels, is sitting in a conference room at the center's new headquarters, a modern building tucked along the Rue de l'Arbre Bénit (Street of the Blessed Tree). "The staff was basically parachuted in," he says, referring to the Kailahun treatment center where Khan was placed. "Think about what it's like for them. They may not know the difference between a baby in a bed, a pregnant woman in a bed, and a clinician in a bed. Their baseline view of the world is that all of our patients are equal."

Bertrand Draguez has red hair and hazel eyes and a youthful face dusted with freckles. He wears jeans and sneakers, and he has an informal, modest manner. The walls of the meeting room are made of pressed chipboard, the same cheap material that's used in the construction of the organization's medical treatment units. In 1999, Doctors Without Borders was awarded the Nobel Peace Prize for its humanitarian work. The organization delivers medical care to people in crisis—in conflict zones, and in places where a hidden emergency is occurring but the world doesn't notice. The organization is sustained by small donations, and it spends around $1.4 billion a year in funds. On this day, the halls and offices of the headquarters were piled with cardboard boxes. The staff members were just moving into their new offices.

"Think about the staff at Kailahun," Dr. Draguez continues. "They don't know Dr. Khan; they've never been to Kenema. Khan's colleagues from Kenema immediately slot him into a higher level of care. For them it's a no-brainer. But when you go into another microculture"—the culture of Doctors Without Borders—"it may be another story. We also shouldn't forget that an ongoing outbreak may destroy your ability for reflection."

The camp managers were isolated, under siege, not sleeping, beyond exhausted, traumatized by all the deaths occurring in front of them, at risk of being infected, and they felt exposed to danger of attack from the local population. They were making life-or-death decisions in the fog of a virus war.

On Friday, July 25, the International SOS jet for Khan landed at the international airport in Freetown. It parked near the terminal, waiting for him to be brought on board. By now, Khan had developed diarrhea and was vomiting. When officials at SOS learned that Khan had these symptoms, they said the plane wasn't equipped to manage a seriously ill Ebola patient, and Khan would be a danger to the crew. Tim O'Dempsey began negotiating by phone with SOS, along with WHO officials in Geneva, and the camp manager Anja Wolz—all of them trying to persuade SOS to take Khan. But the company remained firm: Khan was too sick to be allowed on the plane. Sometime during the day, camp managers informed Khan that he would not be evacuated to Europe, and that, therefore, he would not have access to experimental drug treatments. By then, they had made a final decision that they would not inform him of the ZMapp in the camp freezer.

That afternoon, while the jet sat at Freetown, Khan's brother Sahid started making calls to Doctors Without Borders. He had been trying to reach Humarr for days. The family was frantic with worry about him. They knew he was in the camp at Kailahun, but they had no idea what was happening or even if he was alive. Sahid eventually reached someone at the Kailahun camp, and he demanded to speak with his brother.

"He's too tired to talk," the person who answered the phone said.

Sahid thought his brother was dead. He kept calling around, and finally he reached Michael Gbakie. "What is your name?" Sahid demanded. "What is your location?"

Michael explained that he was Khan's deputy and he was at the camp.

"Nobody's telling me anything!" Sahid burst out. "How is he? What is our plan?"

Michael explained that a medical jet had arrived in Freetown. There was some sort of holdup, but the government of Sierra Leone was still trying to fly Khan out, possibly with a different medevac company. "The minister of health told me that they are going all out for aircraft."

"Is he alive?" Sahid said. "I want a picture of him to prove he is still alive."

Michael promised to take a picture of Humarr, and signed off. He then went to the visitors' area next to the red zone, where he found Khan sitting in a plastic chair by the fence. Michael snapped a picture and texted it to Sahid.

In the photograph, Khan is sitting slumped on the chair. His face is swollen, his eyelids are heavy and puffy, his expression is masklike. He appears to be exhausted and turned inward, though a slight smile flickers on his face. Sahid now thinks his brother's smile may have been for the sake of their mother, an attempt to tell her not to worry.

FATHER AND DAUGHTER

FREDERICK, MARYLAND
5:30 p.m., Friday, July 25, 2014

Hours after Michael Gbakie took the picture, Lisa Hensley was in her hotel room in Monrovia, where it was 9:30 at night. It was pouring rain. She was looking at James on the screen of her laptop. The Skype call had gone through. James's face was jaggy and freezy, but at least she could see him. Her mother, Karen Hensley, was sitting by James.

"Grandma's driving me crazy," James said.

"What's she doing?" Lisa asked.

"She's telling me *no* all the time."

"Well, James is driving me nuts," Grandma said good-humoredly. "Too many requests for snack food."

Grandpa never said *no*, James insisted. Grandpa was making him chocolate chip pancakes for breakfast with maple-flavored sausages and lemonade to drink. Lisa thought her father's breakfasts were a little questionable, but she didn't say anything. She asked James how his day had gone at summer camp. She reminded him she'd be home soon—her tour of duty was getting close to its end. She told him she'd see him soon, and said goodbye.

Immediately afterward, Hensley took up her U.S. government secure cellphone and called her father. Mike Hensley was in another room in the house, in a place where James and Grandma couldn't hear him. Lisa and her father began talking, and their conversation was dead serious.

Mike Hensley, MD, PhD, is an expert in clinical trials and the licensing of vaccines and drugs. Working for the company that is now called Sanofi Pasteur, he had brought several childhood vaccines through clinical trials and licensing for human use. He had also participated in a clinical trial of an experimental antibody drug for cancer: He knew something about antibody drugs. In mid-July, Mike Hensley had begun focusing on the fact that healthcare workers in West Africa were dying of Ebola, and there was no vaccine or drug to protect them. He got involved with an effort to set up clinical trials in West Africa of Ebola drugs and vaccines in order to rush them into licensing.

Mike Hensley discussed the matter with Pardis Sabeti at Harvard, and he brought in an expert on regulations of clinical trials. He got to know the three inventors of ZMapp, Gary Kobinger, Gene Olinger, and Larry Zeitlin. By the evening of July 25, Mike Hensley had come to believe that the two best options for clinical trials were the VSV-ZEBOV vaccine and ZMapp. He and Lisa had started having twice-daily phone meetings about the project.

Now, Mike and Lisa had a swift, businesslike exchange of information. There was no emotion in their voices. If you had heard the call, you would have assumed it was merely two scientists talking with each other, and you would never have known it was a father and his daughter.

Mike told Lisa that Larry Zeitlin was sending him a large package of documentation on ZMapp. He intended to incorporate it into a plan for a clinical trial of ZMapp in Africa with Africans. He told her about some of the plan's fine points.

Lisa gave her father a report on the situation in Monrovia. The hospitals were filled with Ebola patients. More patients kept coming to the hospitals, and the hospitals were now turning people away even when they obviously had Ebola. Dead bodies had started to show up on the streets. Civil order was starting to break down.

They signed off. Neither of them said "I love you" at the end of the call; it was something understood. "Stay safe," Mike said to his colleague and daughter.

If Lisa broke with Ebola, her colleagues would do everything they could to get her evacuated to the United States. But their efforts could easily fail. The U.S. government didn't have any plans for evacuating personnel who came down with Ebola. If an air evacuation couldn't be arranged quickly, Lisa could end up trapped in a horrifying Ebola ward, with no medical care. Mike Hensley wanted a trial of ZMapp started in West Africa for humanitarian reasons. At the same time, though, he hoped that a trial of ZMapp would get the drug positioned somewhere near Lisa in case she needed it. He said nothing to Lisa, her mother, or to the inventors of ZMapp about this thought; he kept it entirely to himself. "I wanted some kind of treatment on the ground for her," he later said.

MONROVIA, LIBERIA
6 a.m., Saturday, July 26

It was still raining when Hensley got up the next morning, but she wore James's sun hat to the breakfast table. The hat was a hope for sunshine. The lab wasn't staffed that day—it was July 26, Liberian Independence Day, a national holiday. Nevertheless, Hensley, Randal Schoepp, and a third American scientist took an Embassy car to the national lab, where many blood samples were waiting for testing. Among them was a second blood sample taken from the patient Tamba Snell at Samaritan's Purse, who had tested negative the day before. There was also a tube of blood from a patient at Samaritan's Purse named Nancy Johnson. Soon Hensley got an email from someone at Samaritan's Purse: "Tamba Snell" and "Nancy Johnson" were members of their medical staff.

They started with the blood of Tamba Snell, and found he had Ebola. In early afternoon Hensley sent an email to Samaritan's Purse:

"I am very sorry to inform you that Tamba Snell is positive." Later that afternoon the Nancy Johnson blood also came up positive.

At ELWA, Lance Plyler went to the house where Kent Brantly was isolated, in bed, and was distressed to see how ill he looked. "I hate to tell you that you have Ebola," he said. After a moment, Brantly said, "I really did not want you to say that." Plyler immediately decided that he would do all he could. He knew that there were experimental drugs for Ebola. Doctors from Samaritan's Purse sent an email to the top CDC official stationed in Monrovia, a doctor named Kevin M. de Cock. They told him that they wanted to talk to a researcher with direct experience in the development of the drugs. They wanted that person to put Lance Plyler in touch with anyone who might have access to these possible therapies.

After she finished up at the lab, Hensley dropped by a party at the house of an Embassy official. She stayed away from alcohol and watched her phone. Around eight o'clock she got an email from the CDC official Kevin de Cock. He said that Samaritan's Purse wanted to talk with a scientist who had direct experience with the development of experimental drugs for Ebola. Samaritan's Purse wanted to speak with just one person—only one person—about the drugs. De Cock asked Hensley if she'd be willing to serve as the advisor to Samaritan's Purse.

Minutes later, Hensley was riding in an Embassy Land Cruiser on its way to ELWA Hospital. The driver went fast along dark streets. Security was deteriorating in the city, and an attack or a hijacking wasn't out of the question. The driver was under orders to not allow Hensley to step out of the vehicle unless there were security guards present at the location.

The driver turned off the road and stopped at the entrance of the grounds of ELWA Hospital. The area was an empty field of grass and weeds, and was poorly lit. There were no security personnel visible. A pickup truck was parked near the gate, its headlights shining. Hensley's driver went into a three pointer and positioned the Embassy vehicle for an escape.

The truck's door opened and a white man stepped out. He had a gaunt, unshaven face, dark hair, high cheekbones, and a mustache with a soul patch. Hensley's driver didn't like the look of the man, and told Hensley to stay inside the vehicle. After hesitating, she opened the door and got out.

MILKY WAY

The gaunt man was Dr. Lance Plyler, the head of emergency medical operations for Samaritan's Purse in Liberia. Hensley climbed into his truck, and he drove through the compound while the Embassy driver followed Plyler's truck. They arrived at a small house, where a lighted window was opened just a crack. Kent Brantly was sitting in bed behind the window with his laptop computer. He was researching his case, and he knew about antibodies to Ebola.

Standing close to the crack in the window, Hensley quickly summarized nineteen possible options for him to consider. This was swift technical talk between a scientist and a doctor who needed a drug to save the life of a colleague and himself. Hensley had brought with her a spreadsheet, and she went through the list. She had done lab research toward the development of most of the compounds, and most of them had never been tested in humans. In January, Tekmira Pharmaceuticals had begun testing a drug, TKM-Ebola, in humans, evaluating it for safety. It had shown decent results in monkeys, but the drug had been put on partial hold while the company collected more information for the Food and Drug Administration. There was a drug called T-705, which had been tested in Japan, in humans, against influenza virus; the

CRISIS IN THE RED ZONE

drug might have some effect on Ebola. Hensley told Brantly that she
had participated in a study of a drug called rNAPc2, an anticoagulant
made by a company called Nuvelo that had saved one of three monkeys it was tested on. There was the vaccine called VSV-ZEBOV,
which Hensley had also worked on. Another vaccine known as an ad-
enovirus plus IFN-Alpha. A drug called PMOPlusR. And there were
more drugs to consider, as well.

Brantly focused his attention on ZMapp. The drug had saved monkeys. But, still, he didn't know what to do. When Hensley finished her
review of the possibilities, Brantly's voice came through the window:
"What would you do, Lisa?"

She couldn't tell him what to do. She had worked on many of
the compounds, and the drugs were unlicensed and untested. She
was bound by law and ethics not to advise anybody to take any drug
under those circumstances. "These are all very personal decisions,"
she said.

Then she told him that she had been exposed to Ebola, sixteen
years earlier. At the age of twenty-six, working in a space suit with
liquids full of Ebola particles, she had cut her finger with scissors,
which had gone through two layers of gloves. The only experimental
treatment at that time was a horse serum made by the Russians, which
could kill her, and she had decided not to use it unless she was certain
that she had contracted Ebola. On the night of the accident, after a
meeting to analyze what had happened, she was sent home to her apartment. She called her parents and told them that she might come down
with Ebola and that they would have to collect her belongings and take
her cat home with them.*

Brantly listened, and said that of the drugs available he probably
would choose ZMapp for himself, based on the data, even though it
had never been tested on humans. Hensley offered to donate blood if
he had hemorrhages.

Lance Plyler then drove her across the compound to the house of
"Nancy Johnson," who was a woman named Nancy Writebol. Her

---

* The incident is described more fully in *The Demon in the Freezer* (2002).

house was near the beach. She had been working in the disinfection line outside the Ebola ward, spraying people with bleach as they exited and helping them remove their gear. When Hensley and Plyler arrived, Nancy's husband, David, who was considerably older than Nancy, was getting ready to enter the house to help his wife. The house had become a red zone. As he got himself dressed in PPE, Hensley saw that he was fumbling with his equipment. He clearly wasn't a medical person.

He fitted on his mask and goggles, and went inside, and Hensley stepped up to a window to observe. The window was wide open and had a screen in it. Nancy was lying in a bed by the window, and a ceiling fan was turning, blowing air around her to cool her skin. She had a fever and was extremely ill: Hensley could see that she was dying, and that David Writebol knew it.

Nancy wanted to use the bathroom. David helped her out of bed. She had difficulty standing. Then the couple walked slowly toward the bathroom.

Hensley began to feel uncomfortable watching this. She turned her back, to give them privacy, and found herself looking up at the sky. The rain had cleared off, the clouds had vanished, and the sky was a black dome glittering with stars. The Milky Way stretched across the top of the sky in a band of haze embedded with blue-white and golden stars and rifts of dark smoke. Hensley had a moment to think, and she began assessing her life and its meaning.

She thought about what she had just seen. As David Writebol was putting on PPE and preparing to go into the house to help Nancy, he was clearly nervous, frazzled, uneasy, but when he walked into that room all that mattered to him was Nancy.

She thought about the love she had found in her life. For a number of years she'd had an on-and-off relationship with Rafe. He was highly successful, extremely good-looking, fun to be with, and he was a father with children of his own. But she and Rafe had their differences. Weeks earlier, just before she left on her deployment, he had told her that he wanted to take a break from the relationship. It hadn't bothered her that much. Yet there was a romantic part of her that dreamed of living with a man she loved with all her heart, and who loved her the same way. Somehow that storybook hadn't happened for her. It had happened for David and Nancy Writebol.

Maybe, she thought, the love in her storybook would be that of a mother who had raised a child and had worked hard to find medicines that might help some people. *Would Rafe put on PPE and help me out of bed if I was dying? Would he have the courage and love to do that? Would anybody be there to just sit with me and hold my hand as I was leaving this world?*

She pulled herself out of these thoughts. She didn't like dwelling on herself when someone was dying of Ebola. She turned around and faced the window.

By this time, David had helped Nancy back into bed, and she was coughing. Hensley recognized it as the classic Ebola cough. About 35 to 40 percent of Ebola patients have a deep, wet-sounding, unproductive cough known as the Ebola cough. Hensley knew that tiny coughed droplets, invisible to the senses, would be bobbing in the air around Nancy, and the turbulence made by the ceiling fan would be carrying the droplets out through the window and around Lance Plyler and herself. She could smell the air of the sickroom.

MONROVIA, LIBERIA
11:30 p.m., Saturday, July 26

That night, in her hotel room, Hensley sent a text to Lance Plyler. "You guys make me a little bit nervous," she typed, and she advised them to wear breathing masks outside the windows of the two patients.

She counted up her exposures to the virus: there were three. The first exposure had happened when she was talking with Kent Brantly in the storage room and moving boxes around. The room was an enclosed space, where the air was dead and not circulating, and they had been talking face-to-face. Brantly had had Ebola disease then; he had been infective. As a person talks, extremely fine, invisible drops of saliva are sent into the air around the person's mouth, and the drops can drift up to six feet. Had an unseen fleck of the virus landed in her eyes, in her mouth, on her skin?

The second exposure had been when she was standing by the crack in Kent Brantly's window and talking with him. Fine droplets from his mouth, produced when he talked, could have been drifting around her face.

The third exposure had happened while she stood by the open window at Nancy Writebol's house, feeling and smelling currents of air coming out of the bedroom driven by the ceiling fan. Nancy had been coughing.

Three exposures. Not massive, but they were real. What were the chances she'd gotten infected?

She trusted Gary Kobinger, and decided to call him. It was early evening in Winnipeg. She told him about her exposures, and said, "What do you think?"

Ebola researchers have a habit of making morbid jokes about their work. In a teasing sort of way he said, "Do you have a headache yet, Lisa?"

Hensley laughed somewhat uncomfortably. Kobinger went on to say that it didn't sound like anything serious, and he advised her not to worry about it.

After talking with Kobinger about her exposures, Hensley made some decisions. She was obligated by ethics and government rules to report her exposures to the Integrated Research Facility, her employer. That is, she had to inform her boss, Peter Jahrling. She made a mental note to call him.

If she was now infected, she would be in the incubation period, the time during which the virus is replicating, but there would be no symptoms. The typical incubation period of Ebola is about eight days, but the virus can incubate in a person for up to twenty-one days before symptoms appear. Her tour of duty was about to end. If she had the virus in her body she could bring it into the United States and break with symptoms afterward. She could infect her colleagues, the general public, or her son. This was an unacceptable risk.

She decided to extend her tour of duty for an extra two weeks, to cover a reasonable incubation period of Ebola virus. If the virus was in her body she simply couldn't go back home now. And she had made a promise to Samaritan's Purse and to the two Ebola patients to help them through their crisis, including offering her blood to Kent Brantly for a transfusion. She would stay in West Africa, she would stick with the patients, and she would start taking her temperature twice a day.

WINNIPEG AND SAN DIEGO
Night, Saturday, July 26

Gary Kobinger had joked with Lisa Hensley about her exposures and told her they were nothing to worry about. In fact, though, he was alarmed. The exposures showed that Lisa Hensley was starting to take chances. She was putting herself at risk, getting too close to the virus, not paying enough attention to her personal safety. If she continued to take chances the odds of an infection would go up. She might already be infected.

Soon after he finished talking with Hensley, Kobinger got in touch with Larry Zeitlin, the chief executive of Mapp Bio. The two men spent the next several hours in a back-and-forth discussion about how to help Hensley if she broke with Ebola. They knew very well what had happened to Humarr Khan. Hensley could end up in Khan's situation, trapped in a camp or in a chaotic Ebola ward in a Monrovia hospital. Humarr Khan was still alive, but he could not be evacuated, and the Doctors had decided not to offer him ZMapp. If Hensley was sick with Ebola, she would not be allowed to get on a plane. Thus she might not be able to get evacuated, either. If she couldn't get to Switzerland or the United States, someone would have to bring ZMapp to her.

Kobinger and Zeitlin started discussing how to get three doses of ZMapp to Lisa Hensley in case she needed them. There were six official courses of ultra-pure ZMapp in the world. Course No. 1 was sitting in Geneva. Course No. 2 was sitting in the freezer at the Doctors' camp in Kailahun. The course at the Doctors' camp *might* be available for Hensley, except that transferring the drug from Sierra Leone to

Liberia, across an international frontier, would be extremely difficult
due to bureaucratic restrictions. Official courses Nos. 3 through 6 were
sitting in Owensboro, Kentucky, and elsewhere.

In addition, there was that extra *unofficial* course of ZMapp—the
hidden, rainy-day supply. Three little bottles of ZMapp sitting somewhere, frozen, pure, ready for use. This was Course Zero, the secret
stash. Kobinger and Zeitlin worked out a plan.

SURF

As Hensley slept, Zeitlin was at home in San Diego, stretched out on the living room couch, exchanging emails with Kobinger on his laptop. Zeitlin's house was dark and quiet. His wife was upstairs in the bedroom, and their children—a five-year-old and a baby—were asleep. He typed an email to Hensley:

> If you were exposed and god forbid test positive, let us know immediately (our cellphones are on 24 hours/day) and Gary or I will fly with a course from North America to wherever you are. This would likely have to be off the record unless you were able to get to Geneva or the US.
> Best
> LZ

It seems that the two scientists intended to try to get a course from the Kentucky facility, but this would certainly involve a lot of government bureaucracy and paperwork, and Hensley could die. If Hensley was dying, they were ready to take Course Zero out of its hiding place, smuggle it to her, and give it to her themselves, regardless of the consequences. (Hensley herself had nothing to do with this plan.)

It is illegal to transport an experimental, unlicensed, untested drug across international borders with the intent of administering it to a person without any sort of oversight from the governments involved.

"If we got caught bringing ZMapp to Lisa, I certainly didn't know what the consequences would be," Zeitlin now says. "It was certainly illegal. It was a dangerous thing. We were a small company, and this was putting the company at risk. Worst case, it could be jail time for me. And I could be banned from drug development," Zeitlin says.

Whatever the consequences of bringing ZMapp to Lisa Hensley, the two scientists had decided to take the risk if it would increase her chances of survival. Zeitlin's email also made it quite clear that if the drug failed to save Hensley, and she were to leave this world, Larry Zeitlin or Gary Kobinger, or both of them, would be at her bedside holding her hand.

ELWA HOSPITAL, MONROVIA
6 a.m., Sunday, July 27

Dawn broke over West Africa. As the sun came up, Dr. Lance Plyler, the chief of emergency operations at Samaritan's Purse at ELWA Hospital, was sitting in bed in his house by the beach, reading a psalm in the Bible. Through an open window came a smell of salt air and the rush and thud of surf breaking on the sand. Farther out to sea, fishermen in wooden boats were hauling nets.

> You shall not be afraid of the terror by night,
> *Nor* of the arrow *that* flies by day,
> *Nor* of the pestilence that walks in darkness.

The words of the psalm didn't help. Plyler was really afraid of what was coming for Kent Brantly and Nancy Writebol.

He spent the day arranging air evacuation for the patients. Samaritan's Purse eventually hired Phoenix Air, a private air service, to carry the patients to the United States in a specially equipped Gulfstream III medevac jet, which would have in it a biohazard isolation capsule for one patient. The jet could carry only one patient at a time. The patient would be admitted to a high-biosecurity ICU at Emory University Hospital in Atlanta.

Phoenix Air began installing the capsule in the jet, but the work was time-consuming, and the jet wouldn't arrive in Monrovia for three days. It wasn't clear that Nancy or Kent could survive for three more days. While he was arranging for the jet, Lance Plyler also began making phone calls all over the world, trying to find experimental drugs that could be shipped to him in Monrovia. He reached Tekmira Pharmaceuticals, an American company, and they agreed to send him some TKM-Ebola, the company's experimental anti-Ebola drug.

KENEMA

About 10 a.m., Sunday, July 27

Mohamed Yillah, the brother of Auntie Mbalu Fonnie, had shut himself in his bedroom and told his family to leave food for him outside the door. His first blood test for Ebola came up negative. He got a second test, and this time a technician called him and said the result was positive.

After he got the news, Yillah walked out of his bedroom with a smile on his face. "It's all over, I've tested negative," he said to his family. In fact, he was feeling much better, he told them, so he would be returning to work. He would be going on a mission to Kailahun for several weeks, and cellphone service was spotty there. "If you call and don't get me, maybe I have gone off coverage," he said to his family. Sounding casual, and being careful not to touch anybody, he said goodbye to his wife, his children, and to his mother, and got on his motorbike and drove off. They didn't realize he had said farewell.

Yillah parked his bike at the hospital and got into the back of an ambulance, which carried him to the Doctors' camp in Kailahun. Lying next to him in the ambulance was Nurse Alice Kovoma, who had tried to save Mbalu Fonnie and had prepared her body for burial. They ended up in Khan's tent, lying in cots next to him. By then, a Kenema Hot Lab technician named Mohamed Fullah had also been put in Khan's tent. Fullah had been running an informal clinic in Kenema, and it seems he had caught the virus from a patient at his clinic.

Yillah was now very sick. Despite his condition, he began giving care to Khan and to the other Kenema staff in the tent. That Sunday night, Yillah's phone rang. It was President Koroma, calling to offer encouragement to the Kenema medical staff. Yillah told the president that they were all alive by God's grace, and he handed the phone to Khan. Khan spoke briefly with the president, who offered him words of encouragement. Right then, Joseph Fair was working late at the Ministry of Health in Freetown, and somebody told him that President Koroma was on the phone with Khan. After the president had signed off, Fair got on the line. "Cee-baby, it's me, Joseph. How are you doing?"

Khan answered in a hoarse voice, in words spoken in short bursts. He had stopped eating, he said. He couldn't keep anything down.

"Is there anything I can do for you?"

"I really want Pringles and a Sprite."

Fair promised to do what he could.

ELWA HOSPITAL
5 a.m., Monday, July 28

Kent Brantly woke up in darkness, an hour before dawn. He stumbled into the bathroom and sat on the toilet and had three heavy bursts of diarrhea. When he had finished, he stood up and looked. The toilet was full of blood—black hemorrhage. He had just lost between a pint and a quart of blood. A wave of dizziness hit him, and he almost fainted. He steadied himself, and looked in the mirror. His eyes had turned bright red overnight—he was hemorrhaging into the whites of his eyes, a strong predictor of death. He also could see, for the first time, the classic Ebola rash on his chest and torso, a sea of red pimples mixed with small blotchy, starlike bleeds visible below the surface of his skin. He was bleeding into his skin. This, he knew, was also a strong predictor of death.

Later that day, Brantly's fever spiked to 104.5. The rash spread up his neck and down his legs, and he became too weak to get out of bed.

He hemorrhaged into a bedpan. Then he was put in diapers, which got soaked with diarrhea and blood. A doctor gave him a transfusion of whole blood to replace what he was losing. He prayed with his caregivers and thought constantly about his wife, Amber, and their children back home in Fort Worth.

FREDERICK, MARYLAND
Monday morning, July 28

Mike Hensley made pancakes and sausage for James's breakfast, and then drove him to camp at the YMCA. Afterward, he stopped at a Starbucks and sat at an outdoor table, drinking coffee and reading the news on his phone. He saw that two American medical workers in Liberia had gotten infected with Ebola. Just then his phone rang.

It was Lisa. There was a roaring noise on the line, and he couldn't hear her very well. He realized it was the sound of air rushing in her space suit—she was calling from inside the lab.

She needed his advice about antibody drugs when they are delivered by intravenous infusion. Mike Hensley had worked in a trial of an experimental antibody drug for cancer. He had done IV infusions into patients. Lisa wanted to know what could go wrong during such an infusion. Could the patient die of an allergic reaction?

He guessed she was talking about the two Americans sick with Ebola, and he guessed the drug was ZMapp. But he knew she couldn't tell him anything about it because of medical privacy laws. "The outcome is fairly predictable," he said to her.

"WHAT, DAD?" she shouted. She couldn't hear him over the noise of her space suit.

"THE OUTCOME IS FAIRLY PREDICTABLE," he said loudly. He had already been thinking a lot about the effect of ZMapp on Lisa's body if it were given to her by infusion. He did not tell her this. Speaking loudly and distinctly, he said there were two kinds of bad reactions from an antibody drug. The first kind, which was rare, would be an immediate severe allergic reaction. The second kind, much more common, was flu-like symptoms—chills, fever, muscle aches.

She asked if an allergic reaction could be managed.

"YES. THE DOCTORS SHOULD HAVE FIVE MILLI-GRAMS OF BENADRYL AND A HEFTY DOSE OF IV DECA-DRON OR SOLU-CORTEF DRAWN UP AND WAITING."

People at nearby tables were glancing at him.

"What should be done if the patient gets flu-like symptoms?" she asked.

"WITH FLU-LIKE SYMPTOMS, THEY SHOULD STOP THE INFUSION FOR FIFTEEN MINUTES. GIVE THE PA-TIENT IBUPROFEN, AND THEN RESUME ADMINISTER-ING THE DRUG. DON'T STOP THE DRUG EVEN IF THERE'S A SEVERE REACTION, JUST TREAT THE REACTION AND KEEP GOING."

She thanked him and said she had to get back to work.

"BE SAFE," he said loudly.

At ELWA Hospital, Lance Plyler had been contacting drug companies about experimental drugs. He phoned Gary Kobinger in Winnipeg, and asked whether it would be possible to have some ZMapp shipped to him.

Kobinger dropped a bombshell on Plyler during the phone call: There was a course of ZMapp sitting in a freezer in Sierra Leone, at the Doctors' treatment center in Kailahun. Doctors Without Borders had decided not to use the drug.

Samaritan's Purse arranged for a bush pilot to fly a light plane to pick up the drug, but there was no airstrip near the Doctors' camp. The nearest airstrip was in Foya, Liberia. Foya was in chaos. A medical team in Foya had been attacked by local people, refugees were fleeing across the border to Guinea, and U.N. soldiers were being sent in to restore order. The pilot might not be able to land in Foya, or the plane might get trapped on the ground after it landed. The American Ambassador to Liberia, Deborah R. Malac, had been following the crisis at Samaritan's Purse, and she offered the help of the U.S. State Department. Ambassador Malac and her staff then began organizing a backup flight to Foya in case the bush pilot couldn't complete his mission. By late afternoon, July 28, the Embassy had arranged for a United Nations helicopter to fly to Foya and pick up the drug if the bush pilot couldn't get through. Since this would be a U.S. government operation, a government official would have to be on board the helicopter. That evening, a Marine lieutenant colonel at the U.S. Embassy named Bryan Wilson phoned Lisa Hensley and asked if she would be available. Hensley was on a military deployment, and she agreed to go. The helicopter was scheduled to leave Monrovia the next morning at first light.

She still needed to call Peter Jahrling to let him know that she'd had some exposures to Ebola and had decided not to return to the United States until the incubation period had passed. She worked late in the lab testing blood, and at ten o'clock she exited from the lab, got out of her space suit, and called Jahrling on her government phone.

ARGUMENT

MONROVIA AND FREDERICK
10 p.m. West Africa time

Rain was pouring, and Hensley couldn't get a signal on her phone. Tropical storms were moving through, and communications had gone bad. Hoping for a better signal, she stepped outdoors to a rusty iron balcony that looked out over the roofs of the chimpanzee cages. It was pitch dark and pouring rain, but now she got a faint signal, and she reached Peter Jahrling in his office.

Between the poor connection and the rain pounding on the metal roof of the chimp house, Hensley and Jahrling could barely hear each other. She told him she was going to fly on a helicopter to Foya the next morning to pick up a course of ZMapp for possible use in one of the two Americans sick with Ebola.

"You're doing *what*, Lisa?" Jahrling said loudly.

Peter Jahrling almost blew a fuse. Hensley was flying around in helicopters with untested antibody drugs? And she could be infected with Ebola herself right now? "Lisa," he said in a tense voice, "you need to come home. You should be getting your air tickets right now."

"I can't leave. The situation is going to get worse."

"Exactly! It's going to get worse! And you're coming down with Ebola and you want to *stay* in Monrovia?" Jahrling couldn't believe what he was hearing from his deputy. She could end up bleeding out in a tent. He told her that if there was *any* chance she was infected with Ebola, she needed to get herself back to the United States immediately. Because if she tested positive for Ebola in Africa, she would be barred from boarding an airplane out. She would be stuck. "If you were my daughter, Lisa, I would want you home," he said.

She reminded him that there was no plan for how to care for an Ebola patient once they reached the United States. "Until you guys have a plan in place for treatment of Ebola patients, I would rather not come home, Pete. Out of respect for my family and for you. I would never do anything to hurt my family."

Jahrling thought this was nonsense. She was an NIH scientist, she could go to the NIH hospital in Bethesda. It was, arguably, the best hospital in the world, and had a high-biocontainment ICU for a patient infected with a Level 4 virus.

Yes, but if she stayed in Africa, she could continue testing blood, she said.

"Testing a few samples of blood isn't going to help!" he said, exasperated.

The rain was pounding, and Hensley got loud. Testing blood saves lives, she said, because every time you can identify someone who has Ebola, you can isolate that person in a biocontainment ward, and so you can prevent that person from spreading the virus to others and creating more chains of transmission. There was no medical countermeasure to the virus, no modern medical defense. The *only* way to stop the virus was to stop people from giving it to other people, she said.

Jahrling felt that the virus had gone so far out of control that just stopping a few people from spreading it wasn't going to have any real effect on the epidemic. "You brought a water pistol to a forest fire."

"Pete, we can't just pull out. What kind of a message are we sending if we just leave?"

No amount of arguing was going to settle the matter. Jahrling was Hensley's superior in the NIH chain of command. He could have given her a direct order to return home. If he did, she would either have to obey his order or resign from the NIH. He had been her boss for sixteen years, first at USAMRIID and now at the IRF. During that time he had seen her rise from being an inexperienced post-doc to becoming the chief of science at the IRF. In all those years, he had never given Hensley a direct order. Right now, Peter Jahrling feared that a key scientist was about to throw away her life testing blood. He was asking her to leave the field, but he wouldn't command her to do so.

After the call, Jahrling felt angry with Hensley. *She's gonna do whatever the hell she's gonna do,* he said to himself. Afterward, he went around the Integrated Research Facility complaining to people about how stubborn Lisa Hensley was.

MONROVIA
5 a.m., Tuesday, July 29

Hours later, in the dark before dawn, heavy downpours were still falling on Monrovia. An Embassy car took Hensley to Payne Airfield, on the outskirts of the city, where a helicopter with U.N. markings stood motionless in the rain. The airport was socked in with thunderstorms, and conditions were too dangerous for flying. Lieutenant Colonel Bryan Wilson, USMC, was waiting for her. After asking her to take the mission, he had decided to accompany her—to provide peace of mind, he said. Hensley and Wilson sat in a waiting room by the tarmac, hoping for a break in the weather. A few miles away, at ELWA Hospital, Dr. Kent Brantly was awake and vomiting maroon-colored liquid into a basin. It was hemorrhage coming from the lining of his stomach. He would need another blood transfusion soon.

FLIGHT

Hensley and Lt. Col. Wilson waited by the tarmac for three hours while thunderstorms swept the airfield with rain and wind. The downpours eased, and they climbed on board the helicopter. It was an old Russian military chopper with U.N. markings, painted gray and generously equipped with dents. The pilots were breezy Ukrainians in possession of demotic English. Hensley and Wilson sat on benches facing each other, buckled seat harnesses across their chests, and put on hearing protectors. The helicopter lifted off.

Almost immediately the flying conditions went bad. What had seemed to be a break in the storms had only been a lull. The downpours became crushing and opaque, but the pilots pressed onward, the helicopter leaning into walls of rain.

There was a window by Hensley's shoulder. When she turned and looked out, she could see almost nothing except moisture whipping across the glass, but now and then she caught a glimpse of a mountain ridge covered in jungle, slipping by below.

At the moment, she didn't know whether Dr. Humarr Khan was alive or dead. What was certain was that he couldn't get a flight out, and Doctors Without Borders was not going to give him ZMapp. But the drug might help someone.

She dozed off in her harness. After a while she woke up, and she noticed that the lieutenant colonel was awake. "I don't know how you can sleep," he said, seeming a little uneasy. If a Marine seemed uneasy, maybe there was something to be uneasy about. "We've been flying in near-zero visibility," Lt. Col. Wilson remarked.

In this epidemic, everybody was flying in near-zero visibility. Below the helicopter, lost in the rain, Ebola was maneuvering in secret. No drugs or vaccines were known to work against the virus. She was on her way to get one course of an experimental compound in an effort to save one human life.

She sensed there would be consequences as a result of her choice to stay in Africa and go on this helicopter flight. She would deal with them later. The real question, for her, was how her actions would look to James when he was old enough to understand them. If she had left her mission and returned to the United States, James would find out some day. What would he think of her choice, when he was old enough to understand it? Viruses more powerful and dangerous than Ebola were going to emerge in the future, and medical people were going to have to deal with them. "If we don't help, what message are we sending our children?" Hensley would later say. "Our children are going to inherit these problems, and people are dying. Part of the responsibility of a parent is to teach our children how to be responsible. We have to set the example for our staff, our families, and the patients in Africa."

Hensley's helicopter descended into a brushy valley surrounded by the forested hills of Foya, Liberia. As the chopper touched down, she noticed U.N. soldiers wearing combat vests and HEPA masks, and holding assault rifles. She and Lt. Col. Wilson got out of the helicopter and learned that the bush pilot from Samaritan's Purse had landed there an hour before, and had already departed with the drug.

They strapped themselves in, and the chopper carried them back to Monrovia. By the time they landed, the drug had already been delivered to Lance Plyler at ELWA Hospital. Hensley and Wilson stopped at a café in downtown Monrovia for a cup of coffee and a sandwich.

She planned to proceed to ELWA Hospital immediately afterward. It was 1:15 in the afternoon.

KAILAHUN ETC
About 1:00 p.m.

While Hensley and Wilson were sitting at the coffee shop in Monrovia, Michael Gbakie entered the Doctors' camp in Kailahun to check on Humarr Khan. He walked through a zigzag of passages marked by plastic fences, and ended up in the visitors' area adjacent to the red zone. Standing six feet back from the red zone fence, he faced Khan's tent and called out, "Dr. Khan!"

Nothing happened. He kept calling Khan's name, waiting for Khan to come out, as he had always done in the past when Michael shouted to him. Minutes ticked by, while Michael got increasingly worried. There was something wrong. "I decided to go inside so I could make my own assessment," he said later. Standing in the visitors' area, he took out his cellphone and called a doctor at the camp, and asked that he be allowed to go into the red zone with a Doctors worker so that the two of them would check on Khan. The doctor told him this would not be possible.

Michael was starting to get angry. He had the phone number of the minister of health of Sierra Leone, Miatta Kargbo, and he called her and told her about the situation. She, in turn, phoned Anja Wolz, the manager of the Kailahun camp, and told her to allow Michael to enter the red zone so that he could assess Khan's condition.

After the minister called her, Anja Wolz called Michael. By his account, she said, "What is the problem?"

He was pissed. "I want you to allow me to dress and go in to see Dr. Khan. I have not seen him in the visitors' area today."

"According to protocol we do not allow people to go in."

"I am with the Kenema Lassa program. I am very experienced with PPE." He told Wolz that he was now going into the red zone with an IV kit of Ringer's solution for Khan and nobody was going to stop him.

Wolz relented.

Michael went to the dressing area at the camp, where workers were issued cotton scrub suits, impermeable suits, and the rest of a full PPE kit. When somebody handed him a scrub suit, he saw the material was old, and it looked dirty to him. He got mad. "No. I won't have this. I'm an infection prevention and control officer, and I don't know where this scrub suit came from. I have my own scrub suit and boots." He went to his vehicle and got the items, and returned to the dressing line. Then he dressed himself in PPE while a worker gave him instructions on the Doctors' dressing procedures. Then he and a Doctors worker entered the red zone and began making their way toward Khan's tent.

ELWA HOSPITAL, MONROVIA
1:55 p.m., Tuesday, July 29

About the time Michael went into the red zone, an Embassy vehicle dropped Lisa Hensley at the offices of Samaritan's Purse at ELWA Hospital. She climbed a stairway lined with breeze-blocks and found Lance Plyler sitting at his desk, staring at a dented, dirty cube of foam sitting on the floor, which was held together with packing tape. He had a look of panic on his face. "What do I do with this stuff?" he said. He didn't seem to want to touch the cooler.

"Do you want me to open it?" Hensley said.

"Yeah. That would be great."

She peeled off the tape and lifted the lid. A swirl of ice fog ten-drilled out. Sitting in a nest of dry ice at the center of the box were three small plastic bottles. They had screw caps, and the caps had been dipped in wax to seal them. This was Course No. 2 of ZMapp.

Plyler stared at it. "What do I do with this stuff?"

"Let's call Larry Zeitlin," Hensley said. She reached him at home in San Diego, where it was seven o'clock on a Tuesday morning. He was getting his five-year-old dressed while his wife looked after their newborn baby. Hensley handed her phone to Plyler.

"What should I do with this drug?" Plyler asked Zeitlin.

Larry Zeitlin couldn't tell him what to do. As scientists involved with the drug's development, he and Lisa Hensley were barred from giving any advice on whether or not to give the drug to a patient.

Hensley could see Plyler's struggle etched on his face.

Lance Plyler was the lead physician in charge of Nancy Writebol and Kent Brantly. There was only one course of the drug, but there were two patients who needed it. Both were dying. Plyler would have to choose between the patients—offer the drug to one of them and let the other one die. The drug was unlicensed and untested and had never been introduced into a human body. It could kill a person in minutes, especially if the person was already dying. The best decision might be to put the patients in God's hands and keep the drug in the cooler and not give it to anyone.

"There is no right decision and no wrong decision," Hensley said, trying to reassure him. She emphasized that she couldn't guide him to a decision. However, if he *did* choose to give the drug to somebody, she said to him, he should follow a protocol laid out by the drug's principal inventors, Gary Kobinger, Gene Olinger, and Larry Zeitlin. According to the protocol, the three doses must be given to *one* person. He must *not* split the drug between the two patients—not try to stretch it out. If he were to split the drug between two people, the drug wouldn't help either of them and both patients could die.

Furthermore, she told him, the drug should be given to the person who is *less* sick. If ZMapp was given to the sicker patient, it might have a positive effect but the person could still die, and then the drug would have been wasted. In other words, Lance Plyler was supposed to practice triage.

Plyler seemed paralyzed by the weight of the decision. "What should I do with these if we don't use it?" he asked.

"I'll get them back to Larry Zeitlin." She put the lid back on the box, fitting it tightly so that the dry ice inside it wouldn't melt. At about 2:30 p.m. she left Plyler alone in his office, staring at the box on the floor.

KAILAHUN ETC
2 p.m.

Michael Gbakie and a worker from Doctors Without Borders arrived at the small tent where Khan and the other Kenema medical staff had been placed. They found Khan lying on his cot. There was mess all around him of vomitus and stool. His clothing was soaked and filthy, and the pad underneath him hadn't been changed in a while. Michael was wearing goggles and a mask, and he was completely shrouded in hazmat material. "Dr. Khan," he said through his mask. "It's me, Michael."

Khan didn't seem to recognize him.

"It's me, Michael," he shouted.

Khan picked up his head and looked at him.

Michael helped him sit up, and then he took off all of his clothes, cleaned his body with disposable pads, got him dressed in fresh clothes, and put fresh pads under him. As Michael was dressing him, Khan spoke for the first time, and asked for a soft drink. Michael helped him sip it. "I want to rest a little bit," Khan said.

He got Khan adjusted and lying down. He decided to exit and try to find Khan's houseman, Peter Kaima, who had stayed with Khan. He went through a decon line, got sprayed with bleach, and workers helped him remove his gear. He found Peter Kaima just outside the camp. He talked with Kaima for a little while, and then went back to the visitors' area in order to call out to Khan's tent and ask how he was doing.

Meanwhile, Mohamed Yillah, the brother of Mbalu Fonnie, was lying in a cot next to Khan. Yillah had been giving care to Khan, but he hadn't had enough strength to get out of his cot and help clean Khan. Now he did get out of his cot. He thought Khan needed some fresh air. He got Khan sitting up, then lifted his feet out of the cot and placed his feet on the plastic floor of the tent. Then Yillah got his arms around Khan and lifted him out of the cot. Carrying Khan in his arms the way you would carry a baby, Yillah walked out of the tent. There is no medical explanation for how a man in late-stage Ebola disease could pick up another man and carry him.

Staggering, going step by step, Yillah carried Khan to the visitors' area, then laid him down on a soft plastic mat on the ground near the fence. Then Yillah collapsed in a chair next to Khan. After a while, though, Yillah managed to get himself out of the chair, and he went back into the tent and lay down on his cot.

Shortly afterward, Michael entered the visitors' area, and he found Khan lying on the mat by the fence. Khan was gasping for breath. "Dr. Khan?"

Khan didn't respond.

"Dr. Khan!"

Khan turned his head. "Michael. . . ." He seemed to swallow the word *Michael* down his throat, and nothing more came out, and his breathing stopped.

"Dr. Khan is dead!" he shouted.

KAILAHUN ETC
About 3 p.m.

At the moment Michael shouted, "Dr. Khan is dead," Khan's older brother Sahid, who was in Philadelphia, was talking on the phone with Peter Kaima, who was standing just outside the camp. Sahid heard a series of cries erupt in the background. "What is happening?" he asked Kaima.

"Doctor has left us," Kaima answered, and started to cry.

That was when Sahid Khan realized that he had just heard the moment of his brother's death, over an open phone line.

Mohamed Yillah, who was lying in his cot in the tent, having just carried Khan into the open air, felt a wave of regret. He had left Dr. Khan to die alone. He hadn't realized Khan was dying, or he would have stayed with him. At that moment, Yillah wanted more than anything else to be able to rise up from his bed and have a last sight of Khan, but he found himself utterly unable to move.

Minutes afterward, Pardis Sabeti learned that Khan was dead, and she wept uncontrollably. Much later, she reflected on Khan's death. "In the fight with infectious disease we see death all the time, and we wonder why it happens. We're all trying to understand our place in the universe and why we're here. Khan's death left me with a feeling that we just have to do more, and that men like him can't be lost in vain."

KAILAHUN ETC
About 4 p.m.

As the afternoon wore on, Auntie's brother Yillah, who was lying in his cot unable to move, started getting angry. It distressed him that patients in the Kailahun camp weren't getting IV hydration. He felt that he had just seen the result of that policy, in Khan's death. *You need fluid to replace what you are losing. A person who has no fluid is just waiting to die.* The Kenema medical staff still alive in the tent were himself, the lab technician Mohamed Fullah, and Nurse Alice Kovoma.

Yillah wasn't going to take any more of this. His phone battery still had a charge. He called Simbirie Jalloh in Kenema and asked her to send ambulances to the Doctors' camp and bring all the surviving Kenema health workers back to the Kenema hospital. "I want to go back to Kenema to die," he said to her.

Night was coming on, but she promised him that ambulances would leave at dawn the next morning to pick up him and the other Kenema health workers. She had already dispatched one ambulance to pick up Khan's body. Early that evening, Kenema lab technician Fullah died in his cot; afterward there were only two Kenema workers left alive in the tent—Mohamed Yillah and Alice Kovoma.

Camp workers picked up Khan's body on a stretcher and carried it to the morgue tent. Shortly after sunset, workers delivered Khan's body, inside the bags, to Michael Gbakie. He and other workers, all of them wearing PPE, loaded the body into the back of the ambulance, and Michael and a driver set out for Kenema.

The driver went fast on the rugged road. Michael bounced in the passenger seat, staring through the windshield. There was nothing to see: Clouds covered the sky, and there were no lights in the countryside because there was no electric power in that part of Sierra Leone. He thought, *If a doctor we were all listening to has died of Ebola, then what will be our fate? What will be my fate?* The Kambui Hills appeared, their outlines barely visible against the sky, and the sparse lights of the city drew near.

HIDDEN PATH

Dr. Lance Plyler was lying in bed in the dark, looking at the white foam cooler. It sat on the floor next to his bed, a ghostly enigma. The drug inside the cooler could kill Kent or Nancy. Or it could rescue one of them. Or it could do nothing.

Dawn came, and Plyler opened his eyes. The cooler was sitting there. He hadn't been able to move it or even touch it since he'd left it next to his bed. He got up and went into the kitchen.

He was sharing the house with other staff members of Samaritan's Purse. He made coffee for everybody, and went back to bed with a cup. Sipping coffee, he took up his Bible, read Psalms, and prayed. The shipment of TKM-Ebola, from Tekmira Pharmaceuticals, had been lost in transit and hadn't reached Monrovia. The three bottles of ZMapp were the only option now. There was enough of the drug for only one person.

Kent Brantly was a good friend of his. If he offered the drug to Kent and it killed him, he would have killed a friend. Or the drug might save Kent's life. But if he offered the drug to Kent, then he would have to deny the drug to Nancy Writebol, and she would then almost certainly die. Nancy deserved the same love and justice that Kent deserved. But Nancy was sicker than Kent, closer to death. Plyler knew the protocol: He should offer the drug to Kent, who wasn't as sick as Nancy, and he should let Nancy die.

He couldn't forget his Hippocratic oath: "First, do no harm." Every possible choice he could make had the potential to cause deadly harm to at least one person and possibly to two people. How could he make a decision if every decision could cause deadly harm? He prayed for guidance, hoping to receive a sense of what he should do. He felt God near him, but it seemed as though God was guiding him only one step at a time.

Lisa Hensley couldn't advise him what to do. She had participated in the drug's development. And in any case she wasn't a physician. He was the physician. Yet he wanted to talk with her again, not on the phone, face-to-face. He texted her, asking her to come to the hospital.

ELWA HOSPITAL
1:30 p.m., Wednesday, July 30

In early afternoon, she arrived in an Embassy car. She found Plyler alone in his office at the top of the breeze-block stairs, sitting at his desk.

Hensley sat down in a chair facing Plyler.

He was in agony. "I don't know what to do. I just want to save my friend. What should I do?"

"I can't tell you what to do."

"Would you take the antibodies yourself, Lisa?"

She had to pause and think before she answered. If she told him she would take ZMapp for herself, would this be a recommendation that he *should* give the drug to a patient? *You don't want to cross that bright line, the ethical line,* she told herself. After a long pause, she said, "If it was me, I would take them." After the words came out she wondered if she'd said too much.

"Lisa, I'm not trying to question your integrity, but I'm just going to shoot it to you straight. I know you have a conflict of interest. I know you want to help, but you've been developing this drug for years, and you have an interest in giving it to a person." *I have just met this woman,* he thought. *Can she go beyond her scientific endeavors and do the right thing?* "I am begging you, please!" Plyler burst out. "If this was one of your own, a family member, would you give it to them?"

They were staring at each other across his desk, eyes locked. She didn't answer.

He kept his eyes on her, watching her face for a clue that might reveal her emotions or her thinking. He saw her look away, and she seemed to focus her gaze on something that wasn't in the room. *Something private and painful,* he thought. Momentarily he wondered if she had children. Then he remembered that she'd mentioned a child, a boy. He didn't know anything about the boy.

Excruciating moments passed. The moments got longer, and still Hensley didn't speak. *What would happen if I got Ebola, Mom?* If the drug were given to a child with hemophilia, it could be unpredictable and dangerous.

Finally she broke the silence. "Yes. I would give it to a child of mine."

Her simple answer after such a long silence convinced him that she was telling the truth. Even so, he sensed deep emotions in her that he couldn't see or understand. *There is something heavy on her heart,* he thought.

As he searched his own heart, Plyler still couldn't see a path to a choice. He told her that he still didn't know what to do.

At this point, Hensley suggested that they set up a conference call with the three principal inventors of the drug—Gene Olinger, Larry Zeitlin, and Gary Kobinger. None of them answered their cellphones. Then she said to Plyler, "Let's call my dad." She explained that her father was a scientist who had run clinical trials of experimental antibody drugs on human subjects.

Mike Hensley answered immediately. She switched her phone to speakerphone and placed it on Plyler's desk, and they leaned in over the phone until their heads were almost touching.

"Dad, would you give me ZMapp if I had Ebola?"

Mike Hensley answered promptly: "Yes, I would give it to you, Lisa." He had already spent a lot of time thinking about that.

Larry Zeitlin called in.

Plyler questioned Zeitlin: "Would you take ZMapp yourself?"

Zeitlin had to think for a moment. After a pause, he said, "With the caveat that it would be ethically dubious for me to recommend ZMapp, I would take it myself."

Hensley said, "Larry, the issue is, would you give it to your child?"

Plyler was hunched over Hensley's phone, waiting for an answer from Zeitlin.

This is indeed the question, Zeitlin thought. He thought about his five-year-old daughter and his baby. At that moment he felt very unimportant and a long way from Africa. He took a breath. "Yes, I would give it to my child."

Afterward, Plyler decided to visit Kent Brantly to check on his condition and pray with him. He drove his pickup truck to Kent's house and stood at the window and looked in. A doctor named Linda Mobula, dressed in PPE, was caring for him.

Kent was lying motionless in bed. He was conscious and in excruciating pain. His eyes were bright red, his pulse was racing, and his breathing consisted of shallow, rapid panting. The rash now covered his body from head to toe. He had been given three blood transfusions to replace what he was losing in hemorrhages, and his diapers were getting soaked in fluids and blood. Kent didn't think he had the strength to keep breathing for much longer. If he couldn't breathe on his own he would die, because there was no oxygen or respiration equipment at the hospital.

Lance and Kent shared scripture and talked briefly. The medevac jet was due to arrive in two days. Two days is a long time when you're dying of Ebola. Kent thought there was a chance he might be able to keep his breathing going for about forty-eight hours, long enough to get on the jet. As soon as he was on board he would be put on a respirator and given artificial life support. The equipment might keep him alive for the duration of the flight, and if he made it alive to the hospital in Atlanta, he would get advanced care. Nancy, on the other hand, couldn't possibly survive long enough to make it onto the jet. "Give the drug to Nancy," Kent said.

Lance left without telling Kent what he would do.

ELWA HOSPITAL
6:30 a.m., Thursday, July 31

The next morning, at break of day, Lance Plyler was sitting in his bed, drinking coffee and looking at the foam cooler. The Atlantic surf beat on the sand outside his window. He hadn't been able to touch the cooler since he'd put it next to his bed. He took his Bible out of his shoulder bag and opened it. The pages were limp and fragile, darkened with sweat from his hands, and marked everywhere with pen and pencil. He turned to the Book of Esther.

In the story, young Queen Esther, a Jew, is married to the king of Persia. Her uncle, Mordecai, discovers a plot in the king's court to have all the Jews of Persia killed. Mordecai urges Esther to warn the king of the plot and thereby try to save the Jews of Persia. He says to Esther, "Yet who knows whether you have been called to the kingdom for such a time as this." At risk of her life, Esther warns the king and thereby saves the Jews.

It seemed to Dr. Lance Plyler that he might have been called to the kingdom for such a time as this. Unlike Esther, though, he couldn't save everyone with his choice. He would *have* to make a choice; there was no course of action that didn't require a choice. One choice would be to do nothing—not use the drug—and let God choose who would live or die. But he felt that, as with Esther, God seemed to be giving him the responsibility of choice. Kent had urged him to give the drug to Nancy. But if he, Lance Plyler, defied the rules of triage and gave it to Nancy, she could die anyway, and he would be abandoning Kent. He closed the Bible.

He wondered if he could split the drug between the two patients after all. This would be a high-risk maneuver. If he gave one dose to Nancy and one dose to Kent, both patients would need to be flown quickly to Atlanta, no delays, no hang-ups. And there would *have* to be more ZMapp in Atlanta waiting for them. If anything went wrong, both patients would die.

Nancy Writebol's house was just around the corner. Later in the morning, Plyler ended up standing at Writebol's window, looking at her. She had end-stage Ebola disease, and was bleeding into her skin and hemorrhaging from her intestinal tract. She was much older than Kent, less physically fit, and it was clear she could die at any moment. As Plyler looked in through the window, his compassion for her blew away the rules of triage. If he didn't do something, she was going to die. He decided, finally, to give the drug to her. He had made his choice.

Later that morning, Plyler left the cooler on Nancy's porch with instructions for Nancy's attending physician, Dr. Deborah Eisenhut: She was to thaw one of the three bottles and administer it to Nancy intravenously in saline solution. The foam cube sat unopened on the porch for several hours, while the staff prepared Nancy Writebol to become Experiment Number One.

KENEMA GOVERNMENT HOSPITAL
Noon, Thursday, July 31

As the staff at Samaritan's Purse was preparing to give ZMapp to Nancy Writebol, the funeral for Humarr Khan began at Kenema Government Hospital. A crowd of five hundred people had gathered in front of the children's ward, where a white casket rested in a viewing area. This was the same place where the crowd had gathered holding candles and singing, hoping to ward off Wahab the Visioner's prediction that an important doctor would die.

After the service, a group of pallbearers, dressed in bio-hazmat suits, carried the casket to a patch of stony ground in front of the unfinished buildings of the new Lassa ward, not far from Khan's smoking place. Gravediggers had started digging, but were having a hard time making progress. The rocks of Kenema are three billion years old, and they resist change. Hours passed while the gravediggers chipped and hacked at the ground. Finally the crowd dispersed, leaving the gravediggers alone with the casket, still digging. A rain shower moved through.

• • •

Soon after Lance Plyler left the cooler on Nancy Writebol's porch, Samaritan's Purse sent out a worldwide press release announcing that they had given an experimental antibody drug to an American sick with Ebola. According to the press release, the drug had been provided by the National Institutes of Health and by Lisa Hensley, an NIH researcher.

Minutes later, CNN picked up the release from Samaritan's Purse and posted a story about it on its website. During the next hour, emails started arriving in Hensley's inbox, coming from Ebola experts all over the world. They were surprised and upset, and they fired questions at her. *Lisa, what did you just do? . . . You provided experimental antibodies to an American patient? . . . The antibodies came from the NIH? . . . Are you crazy? . . . Who authorized you to provide untested antibodies to a patient?*

Hensley was in her space suit in the lab and didn't see the emails. In fact, the drug hadn't yet been administered to Nancy Writebol. The director of the CDC, Dr. Tom Frieden, saw the news reports and called Dr. Anthony Fauci, the head of NIAID—the NIH institute that runs the Integrated Research Facility at Fort Detrick. Frieden was upset, and he asked Fauci what was going on in Africa with this NIH researcher and this NIH drug. Anthony Fauci was taken quite by surprise. It seems he hadn't known anything about it. If an experimental, unlicensed, untested drug is supplied by the NIH and is provided to a patient by an NIH employee, any decision to give the drug to the patient must be handled by top-level administrators at the NIH and must be overseen and authorized by the Food and Drug Administration. It seemed that Lisa Hensley had broken all the rules. The leaders of the NIH knew next to nothing about her. She was a lower-level scientist who'd been working at the agency for only a few months. The heads of the NIH began asking, in effect, *Who is this Lisa Hensley person and what the hell did she do?* Whoever she was, she seemed to have gone rogue.

Peter Jahrling's immediate superior, an NIH administrator named

Cliff Lane, came down on Jahrling and asked for details. Jahrling had to tell Lane that Hensley had flown on a helicopter to get the antibody drug and that she might be infected with Ebola. Jahrling apologized for not having informed senior management about this. Very suddenly, it looked like Jahrling was in trouble.

Jahrling was told to issue a direct order to Hensley to return to the United States immediately, where she would face an investigation and very likely be fired. The investigation would start as soon as possible, even before she returned to the United States. In a space of two hours, Hensley's career went down in flames.

ELWA HOSPITAL
1 p.m.–6:50 p.m., Thursday, July 31

Dr. Deborah Eisenhut tucked a bottle of frozen ZMapp, the first dose, into the bedding near Nancy Writebol's arm, to let it thaw.

While the bottle was thawing, Peter Jahrling called Lisa Hensley and ordered her home. The NIH had already started an investigation and wanted to see every electronic communication she had sent or received before and during her deployment in Africa—every text message, every email, and records of every phone call. Investigating officials wanted to know exactly where and how she had gotten the drug, what she had done with it, and where her authority for these actions had come from.

Lance Plyler knew nothing about Lisa Hensley's recall to the United States or the investigation into her actions: She decided not to tell him about it. Late in the day, Lance Plyler, having made the decision to give the drug to Nancy, got in his pickup truck and drove to Kent's house to see how he was doing. He arrived at the house a few minutes before sunset. The rains had held off, and the sun was setting through incandescent clouds over the Atlantic. By the light of sunset, he looked through the window at Kent Brantly. What he saw terrified him.

CRASH

Kent Brantly's face was a gray mask. His temperature was 104.7. He was breathing thirty times a minute, in shallow pants, and his blood oxygen was dangerously low. At times his breathing would slow down and almost come to a halt, and then he'd suck a lungful of air and resume panting. This is known as Cheyne-Stokes breathing, and it is a sign of imminent death. Lance Plyler had seen many people die, and he knew the look. So did Kent Brantly, who was trying to force himself to breathe. With no ventilators at the hospital, he would never make it through the night.

Suddenly Lance knew what he would do. As he would explain it later, "God gave me an overwhelming peace, and I decided to split the dose." He would break all the rules. He had already broken triage by deciding to give the drug to Nancy, and now he would divide the drug between the two patients. He would do everything the inventors of the drug had told him not to do. He had made his choice, and this time it was final. "Kent, I'm going to give you the antibodies," he said.

"Okay," Kent answered.

There was a problem. All three doses of the drug were half a mile away, and Kent was now actively dying. Two of the doses were frozen rock hard, sitting inside the foam cooler on Nancy Writebol's porch. And one bottle was in Nancy's room and might already be in her bloodstream.

Plyler got into his truck and drove it wildly half a mile down the road, and skidded to a stop by Nancy Writebol's house. He jumped out, opened the cooler, and took out a bottle of ZMapp. It was frozen solid. He stuffed it in his armpit, held it there for a short while, and took it out. It showed no sign of defrosting. It was a useless lump of ice. Kent was dying.

Lance went to Nancy's window and asked Dr. Deborah Eisenhut to bring Nancy's dose to him. The doctor took the bottle out of Nancy's bedding, sterilized the outside of the bottle with bleach water, put it into triple bags, sterilized the bags, and then handed it to Lance through the front door of the house. He traded the bottle for a solidly frozen bottle from the cooler. The frozen bottle was placed in Nancy's bedding, where it began to slowly thaw.

Plyler jumped into his truck, tucked the bagged, half-frozen bottle into his armpit, and drove at high speed back to Kent's house, praying that the drug would melt in time.

He arrived at Kent's house, ran to the window—Kent was still alive. He took the drug out of his armpit and saw that it had melted. He handed it through the door to Dr. Linda Mobula. The time was 7:20 p.m. The sun had set, and the sky was growing dark.

Simultaneously

Night had almost arrived, and fast-moving clouds darkened the grounds of the Kenema hospital. The gravediggers had finished their work. Humarr Khan's pallbearers, dressed in moon suits, passed straps underneath Khan's casket and lowered it into the grave, leaning backward against the straps to brace themselves against the weight of the casket. A handful of people stood watching. One of them was Nadia Wauquier, the French scientist who had been testing blood, and who had enjoyed smoking cigarettes with Khan as they chatted in her office. She watched the pallbearers take off their biohazard suits and toss them into Khan's grave, the standard gesture of farewell to an Ebola victim. The gravediggers began shoveling dirt back into the hole over the suits. Afterward, Nadia suited up and entered the Hot Lab, and resumed testing blood.

ELWA HOSPITAL
Thirty minutes later, shortly before 8 p.m., July 31

While Lance Plyler watched through the window, Dr. Mobula filled an infusion bag with 750 milliliters of Ringer's solution, broke the bottle's wax seal, unscrewed the top, drew up the drug in a syringe, and injected it into the saline bag. They prayed, and Lance sent a text to Lisa Hensley, telling her that he was splitting the drug between the two patients, and he was about to give one of the doses to Kent Brantly. Nancy would get a dose as soon as it was defrosted. "About to start," he texted.

Dr. Mobula set the infusion to a slow drip, and started it running. The time was 8:00 p.m. Kent Brantly had become Experiment Number One, the first human to receive ZMapp. The drip would run for most of the night while the dose of the drug built up slowly in Brantly's bloodstream. But just a minute or two after the drug began hitting his bloodstream, he began shaking. As an Ebola patient dies, the person's body can convulse. Brantly began to shake violently.

Plyler diagnosed it as a form of shivering called rigors. "That's just the antibodies kicking butt on the virus," he said to Brantly, speaking through the crack in the window.

It may have been shivering, or it may have been the agonal stage of death from Ebola infection, the moment when the person dies with shaking or tremors. Brantly's shivering continued, while Dr. Mobula reported that his temperature had started to drop. In fifteen minutes it fell from 105 down to 100—from fatal range down to a mild fever. The shaking went on for half an hour, gradually decreasing, and finally it stopped. It was now thirty minutes after the drug had started entering his bloodstream. Brantly sat up in bed. Just then, Plyler put his phone up against the screen in the window and took a picture of Brantly—his mouth is open, his eyes are sunken and half closed, but he looks very much alive.

At nine o'clock, Kent said that he wanted to visit the bathroom. He hadn't been able to get out of bed for a day and a half, and had been completely incontinent. He got out of bed and walked to the bathroom, while Dr. Mobula supported him and managed his IV pole. He said he was feeling a little better. Slightly more than an hour had passed since the drip had been started. By then, only about 12 percent of the first dose had gone into his bloodstream.

At 9:09 p.m., while Brantly was in the bathroom, Lance Plyler sent a text to Lisa Hensley: "Honestly, he looks distinctly better already. Is that possible?"

She replied hastily. "Gary [Kobinger] said they see changes [in the treated monkeys] within hours. They may look better but then seem to slip a bit. Yes it is possible."

Lance Plyler stayed by Kent Brantly's window all night, praying with him at intervals, and he watched Brantly get steadily better as the hours went by.

In fact, what had happened to Kent Brantly was a medical miracle. The drug clearly had saved Brantly's life. There are no two ways about it, either. ZMapp had hammered Ebola in Brantly's body. It had started killing the swarm of particles in his body only minutes after the first drops of the drug hit his bloodstream. The idea that any drug could wipe out an Ebola infection in about ninety minutes, or that a drug could rip a person out of an Ebola crash at the very moment the person's body is shaking in death throes, seems like a fictional scene in a screenplay, something that reality would never invent. But there it is. At least in Kent Brantly, who was Experiment Number One, the drug was a true angel's sword, and it tore the heart out of the virus. As this is being written, exactly what ZMapp did to Ebola inside Kent Brantly is still a mystery, but whatever happened, it wasn't pretty for Ebola. In a larger sense, though, the drug opened a window into the future. Drugs like ZMapp could put down a biological weapon or stop any emerging natural virus. ZMapp was a crude version of sharper swords that are already being developed and tested. The drug also—maybe— lived up to Larry Zeitlin's original idea (which he got while he was collecting unemployment) that if you could design a drug that would defeat Ebola, you could pretty much kick the shit out of all kinds of viruses.

At ten o'clock that night, Dr. Deborah Eisenhut placed a needle in Nancy Writebol's hip bone and started an infusion of ZMapp straight into the bone. The veins in her arms had gotten soft and fragile, and an infusion needle would have broken the vein and started a hemorrhage. Soon after the drug started hitting her bloodstream, she got terrible itching in her hands. It was probably an allergic reaction. She did not improve noticeably as the drug built up overnight in her bloodstream, though she remained alive. That in itself may have been something of a miracle.

Lisa Hensley stayed awake in her hotel room, monitoring the developments. When she was sure that both patients had started receiving ZMapp, she sent a text to Lance Plyler telling him she'd been recalled to the United States. She didn't tell him why.

Kent Brantly was carried to Monrovia's international airport in the back of a pickup truck and was placed inside the biocontainment capsule in the Phoenix Air jet, and it took off. When it landed in Atlanta, Brantly walked off the jet, wearing a moon suit, and he was rushed by ambulance to Emory University Hospital and placed in a high-level biocontainment ICU. Three bottles of ZMapp had been delivered to the hospital, having been rushed there from Kentucky BioProcessing. As soon as Brantly was inside the biocontainment ICU, a staff of four infectious disease doctors and twenty-one nurses started working on him. ZMapp may have saved his life in a few minutes, but he wasn't well yet, and his life wasn't going to stay saved without an excellent medical team and the world's best medical technology. He arrived at Emory Hospital on August 2.

TERROR

Humarr Khan had been dead for four days. By now, eight of the Ebola nurses were dead, and the surviving nurses were traumatized. Most couldn't go into the Ebola wards anymore, but there were sixty to seventy Ebola patients distributed among the wards, and the number of patients was growing. Some Kenema staff did continue to work, including Nurse Nancy Yoko, who had prepared Auntie's body for burial.

The World Health Organization continued to send doctors to Kenema to try to stabilize the hospital. One of them was John Schieffelin, a pediatrician at Tulane University School of Medicine. He signed up for a three-week tour of duty at Kenema Government Hospital. His pay was one dollar, with twenty-five cents deducted from it to cover administrative costs. Schieffelin had never seen an Ebola patient, had never worn PPE. A Land Cruiser dropped him at the hospital, and he stood in front of the Annexe ward believing that there was a good chance he would never see his family again.

A British doctor named Catherine Houlihan briefed Schieffelin, showed him how to put on PPE, and together they walked into an Omaha Beach of medicine.

The ward was in horrifying chaos. It officially had seventeen beds, but there were about thirty Ebola patients in the ward. There were whole families with Ebola in the ward. Disoriented, infected patients were wandering about. Patients were moving themselves from bed to bed, choosing to lie in beds that seemed cleaner. Sick parents came into the ward, bringing their uninfected children with them, because villages were refusing to take in the children of Ebola-infected parents. Schieffelin and the other WHO doctors didn't know what to do with these healthy children, and they put them in wards where people were less sick and hadn't yet tested positive for Ebola. It wasn't the best thing for the children, but Schieffelin's only other choice was to put them in wards where patients were in late-stage Ebola and were even more infective. "Did we make mistakes? Absolutely, we did. We were just trying to survive, trying to do the best we could, and we were treading water," he said.

Nurse Nancy Yoko worked in the ward for as many hours as she could, but at night there was often no medical staff in the ward. In the mornings Nancy Yoko and the WHO doctors would remove a few corpses, often from the toilets, and leave them by the Ebola ward. Not long after Schieffelin arrived, the Kenema wards had a hundred Ebola patients in them, and Schieffelin and his colleagues thought about closing down the Kenema Ebola wards, to quell the chaos. They realized, though, that if the wards were shut down, infected people would end up at home being cared for by family members, and the virus would continue to spread and more people would die. They had to keep the Kenema wards open in order to draw Ebola-infected people out of the community and keep them in one place.

John Schieffelin had previously worked in the Lassa research program as a pediatrician, and he knew many of the Ebola nurses. When he arrived, he found his friends Mohamed Yillah and Nurse Alice Kovoma in the Ebola ward—they had been carried by ambulance from the Doctors' camp to the Kenema hospital. On the day he arrived, he examined Yillah and saw that his case was fatal: He had hiccups and he was urinating and defecating blood. Schieffelin broke the rules of triage and did everything he could to help Yillah, even though there was no hope. In the same way, he worked on Alice Kovoma, though there was little hope for her. She would die in his care. Mainly, though, Schieffelin focused his efforts on children and young adults. He was a pediatrician, after all. "We all had one or two patients that we took care of when they first arrived," Schieffelin said. "For whatever reason, we put our heart and soul into those patients. Most of them didn't make it."

Rob Fowler, a Canadian WHO doctor, was working in the Kenema wards, covering thirty to forty Ebola patients by himself. "As soon as I walked into my treatment area in the morning, I met people in bed calling out for things," Fowler recalled. "I was asking myself, 'Whose bed do I go to first?' To the three-year-old who's comatose? To the thirty-year-old woman who's speaking the loudest? If I go to someone's bed first, five other people are saying, *'Doctor, Doctor! Please!'* If I hang a liter of fluid, six other people are asking for the same thing. I have a hard time talking about it."

Schieffelin began to notice that the Ebola patients in the Kenema wards were forming themselves into a kind of community. With very few doctors and nurses around to help them, they began to help themselves. The community of Ebola patients developed leaders, younger people who had survived and were recovering. They began helping out with nursing tasks on the ward.

John Schieffelin had been caring for Mohamed Yillah, and to his complete surprise he perceived that Yillah was getting better. His hemorrhaging stopped, his hiccups stopped, and his fever tapered off. Somehow his immune system had fought off the virus. Yillah's survival is a mystery—he hadn't worn PPE while he cared for Auntie, and he'd had massive exposures to the virus. On August 9 the Hot Lab reported that Yillah's blood was negative for Ebola, and he went home to the family compound on the slopes of the Kambui Hills. Yillah had always been a thin man, and now he was a human skeleton. He walked into his house with a smile, and said to his mother, Kadie, "It's all over, I've tested negative."

She didn't believe it. This was exactly what he'd said to her the last time, those exact same words, that it was all over, he'd tested negative, just before he'd gone off to the Doctors' camp to die.

He wrapped his arms around his mother to prove he was negative. She knew he wouldn't do that if he were positive.

A few months later, Mohamed Yillah and I were sitting in a quiet spot on the grounds of Kenema Government Hospital. He was an extremely thin man, well over six feet tall, with a restrained, thoughtful manner. He seemed wrapped in a dreamlike haze of trauma. He looked to be in his seventies. He was forty-seven. He said that he couldn't remember some of the horror he'd experienced. He spoke of his regrets when he'd left Dr. Khan alone to die. "It is very terrible when I can't even remember what it was like. With God's grace my life was saved," he said.

CONAKRY, FREETOWN, MONROVIA, LAGOS
August 2014

As it turned out, the destruction of Kenema Government Hospital was only a beginning flare of the emerging virus, an early burst of what became a viral crown fire in the human species. As the wards in Kenema burned with the virus, the real epidemic began, and the cities of West Africa caught fire with the A82V Makona Variant of Zaire Ebola—the Makona strain. On August 8, the WHO reported there'd been a total of 1,779 cases of Ebola, with 961 deaths.

The virus had reached Nigeria on July 20, traveling in an American lawyer named Patrick Sawyer. He caught Ebola in Liberia from his sister, then flew to Lagos, the capital of Nigeria, on his way to a conference in Calabar, Nigeria. He got off the plane at the international airport in Lagos feeling extremely sick, and ended up at a hospital in Lagos called First Consultants Medical Centre. The hospital's chief physician, Dr. Stella Adadevoh, suspected he had Ebola, and she kept him in the hospital for tests, although Sawyer wanted to leave. The tests confirmed he did have Ebola, and he died soon afterward. Twenty other people contracted Ebola from his case, including Dr. Adadevoh. The virus threatened to go out of control in Lagos, which has a population of twenty million, and many of the city's inhabitants are poor and live in crowded slums and don't have access to medical care. If the virus were to amplify itself in the urban population of Lagos, the city could erupt in a viral equivalent of a nuclear detonation. Sawyer had close contacts with seventy people while he was dying in Nigeria, any of whom could have caught the virus from him and spread it to others. Fast, decisive action by Nigerian health authorities and foreign doctors managed to break the chains of infection that had started traveling out of Patrick Sawyer.

Dr. Stella Adadevoh, who had prevented Sawyer from leaving the hospital, died of Ebola afterward; she is now regarded as a national hero for stopping the virus from spreading more widely in Nigeria.

If Ebola had blown up in Lagos, and one person harboring the virus had then traveled from Lagos to, say, a supercity like Dhaka, Bangladesh, or Mumbai, India, the virus could have done real damage in those cities, and would have had many opportunities to mutate again, further adapting itself to humans. The wild Ebola that had jumped into the little boy in Meliandou had changed into the Makona strain after passing through a few human bodies. If the virus swarm passed through many more people, in long chains of infection, the chances were that more mutations would happen in the swarm, and the school of fish would change again, and the virus could become even more humanized. Ebola was able to change, it was reacting to its new host, and it was starting to travel in human bodies to more distant points on the planet.

Two weeks after Humarr Khan died, and when it was clear that ZMapp had helped to save the lives of Kent Brantly and Nancy Writebol, *The New York Times* published a story about the decision to withhold ZMapp from Khan: "The treatment team, from Doctors Without Borders and the World Health Organization, agonized through the night and ultimately decided not to try the drug." The article also reported that Khan hadn't been offered a choice of whether to take the drug or not. When the article appeared, Robert Garry, Erica Saphire, Lina

Moses, and Pardis Sabeti were very surprised to learn that Khan hadn't been given ZMapp. They had assumed that Khan had been given the drug and that it had failed to save him.

Pardis Sabeti was furious about the decision. She didn't speak publicly about her feelings, but at the Broad Institute her colleagues heard her cursing profusely, in a loud voice, as she moved around the offices near the Ebola War Room. Humarr Khan had been a member of her team and a dear friend, and he had been denied a drug that could have saved his life.

Dan Bausch was quoted in the *Times* article as saying he disagreed with the decision by Doctors Without Borders, and said he thought that Khan should have been asked for his own opinion. "Dr. Khan was the perfect patient, I think, to understand the complexities of that gray area," he said. He also said it was a close call and that he respected the decision of the doctors on the ground. None of the scientists, doctors, officials, or camp managers knew that Khan had been reading up on the experimental anti-Ebola drugs and vaccines, was familiar with the data on them, and considered ZMapp to be his first choice for treatment. It is being reported here for the first time.

The Kailahun camp managers have been reluctant to discuss publicly the reasons for their decision not to offer the drug to Khan. I learned from three different physicians with Doctors Without Borders that members of the Kailahun team had been deeply traumatized, and wouldn't even talk about their experiences privately with other members of Doctors Without Borders. Eventually Anja Wolz, who had been the clinical manager of the Kailahun Ebola Treatment Center, agreed to speak with me; I reached her on the phone while she was at the Brussels center of Doctors Without Borders.

"It has been quite difficult for me to think about it," she said. "Our fear was that if we gave the medication to Dr. Khan we wouldn't know the outcome. ZMapp had never been tested in a human, and it meant we would use Dr. Khan as a guinea pig." She had talked with Gary Kobinger, asking him for his thoughts about the drug. Since he had helped develop the drug, he couldn't advise Anja Wolz to give it to Khan. "Gary said, 'It's for you in the field to decide. You can decide what is best for you and your team.' " Kobinger had offered to relieve her of the ethical responsibility for the decision. "If you want, we can decide for you," Kobinger said to Wolz, meaning that an international panel of experts would weigh in with a recommendation. Wolz told Kobinger that she would assume moral responsibility for the choice.

Her father called her constantly, trying to encourage her. She said to him, "Papf, people are dying and we can't do anything about it." It was impossible to explain to anyone what it felt like to work in an Ebola treatment center, when there were no treatments, when children and teenagers were dying, alone, without any family around them. She *had* to consider the danger of violence if Khan died; there was violence happening already in areas not far from the camp; the lives of all the patients and staff were in her hands. She had once given up cigarettes but now she was smoking them constantly. She kept hearing promises that the SOS medevac jet would soon take Khan to Switzerland, where he could be given ZMapp without endangering lives at the camp. In the end, all the promises were worthless and SOS refused to take Khan on board. Not long after Khan died at the camp, Wolz learned that two Americans had apparently been saved by the same course of ZMapp that she and her team had decided not to offer to Khan. "I felt, oh, oh, have we done wrong? Now, knowing that ZMapp works, it was probably the wrong decision. It was the best I could do given the facts we had at the time, and I still stand by my position. It was something so emotional, so difficult. There were a lot of things I tried to forget for a long time." Her voice faltered and broke, and she began to weep.

Through the late summer of 2014, Pardis Sabeti and her group continued to read the genomes of the Ebola swarm, and would publish the data in real time on the website of the National Center for Biotechnology Information, so that scientists anywhere in the world could see the results immediately. Then, in late August, Sabeti's group published a paper in *Science* detailing their results. They had sequenced the RNA code of the Ebolas in the blood of seventy-eight people who lived in and around Kenema during three weeks in May and June, just as the virus was starting chains of infection in Sierra Leone. The team had run vast amounts of Ebola code through the machines and had come up with around two hundred thousand individual snapshots of the virus in the blood of those seventy-eight people, and they had watched the virus change as it entered the human species.

Sabeti's group also found that the virus had started in exactly one person. As it spread from this first person to the next, and to the next, the swarm mutated steadily, its code shifting as it explored the human species. As the virus jumped from person to person, about half the time it had a mutation in it. Most of the mutations didn't change the proteins in the virus, but every now and then one did, and the virus became slightly different. By the time the virus reached Sierra Leone and got into the bodies of the women who had attended the funeral of Menindor, the faith healer, the virus had already mutated into two genetically distinct swarms. Both lineages of the virus moved out of the funeral of Menindor and across Sierra Leone. But only one of the two strains at Menindor's funeral ended up infecting most of the victims in West Africa. This was the Makona strain, the hot Makona, the dominating mutant.

By September, Pardis Sabeti could see the Makona strain in operation, but she didn't yet know whether there was anything truly different about it. Was there something unusual about the Makona strain? Was it more deadly in humans, or more infective, or both? Why was the Makona strain sweeping through West Africa when the other strains were fading away? She still didn't have an answer to that last question—she still didn't have visibility into the character of the Makona strain. She could read all the letters of code in the strain, but she couldn't yet understand the meaning of those letters.

Some of Ebola's mutations had made the virus less visible in tests. "It shows that you can analyze Ebola in real time," Sabeti said to me, in mid-September. "This virus is not a single entity. Now we have an entry into what the virus is doing, and now we can recognize what we are battling at every point in time." The *Science* paper included five authors who died of Ebola, including Humarr Khan, Ebola ward supervisor Auntie Mbalu Fonnie, and senior nurses Alex Moigboi and Alice Kovoma. Many other members of the Kenema team were coauthors of the *Science* paper, including Michael Gbakie and Lansana Kanneh, who had risked and nearly lost their lives exploring the Makona Triangle for people infected with Ebola. "There are lifetimes in that paper," Sabeti said.

CURED

Kent Brantly made it alive to Emory University Hospital, but he was still very sick. His illness seemed comparable to what Hensley had said happens to monkeys—"They may look better but then seem to slip a bit." The drug had hammered down the virus in his body but certainly hadn't eliminated it. Brantly was given two more doses of ZMapp—this was Course No. 3, flown in from Kentucky. He continued to improve; he was also getting world-class medical care by his team. Brantly's wife, Amber, had arrived, and they were able to talk with each other through a glass window; he had to remain inside the biocontainment unit.

Nancy Writebol had received her first dose of ZMapp at ELWA Hospital. Afterward she remained in bed in her house, being cared for by Samaritan's Purse doctors and staff. She stayed alive. A few days later, Samaritan's Purse doctors administered a second dose of ZMapp to her. At this point all three doses of Course No. 2—the course that might have been given to Humarr Khan—had been used up. Nancy Writebol was still at ELWA Hospital and needed a third dose.

After the Phoenix Air jet had carried Kent Brantly to Atlanta, it turned around and flew back to Liberia, where it picked up Nancy

Writebol and carried her to Atlanta as well. She, too, was placed in the biocontainment ICU at Emory and was tended by the Emory team. There, she received her third dose of ZMapp—this was the last of the three doses of Course No. 3.

Nancy Writebol had a rough time as she fought Ebola in the Emory unit. Her recovery went slowly, but on August 19 she was discharged from the hospital and went home with her husband, David. She requested privacy and didn't want media attention. She later said that she couldn't remember much about her illness. Ebola can cause amnesia, and many Ebola survivors remember little or nothing of their time spent in the embrace of the virus.

Kent Brantly made steady improvement. But as long as the virus was in his bloodstream it was present in the United States. There was always a danger, no matter how small, that the virus could escape from Brantly's body. For that reason, the Centers for Disease Control monitored his blood tests. On August 20, Brantly was declared free of the virus, and he walked out of the biocontainment unit at Emory, and he and his wife wrapped their arms around each other for the first time in months. The next day he was discharged from the hospital. He was thin, but he was on his feet and smiling. He walked through two rows of medical staff giving him applause, and he went home to Fort Worth. Kent Brantly and Nancy Writebol credit their survival to outstanding care by their doctors, to ZMapp, and to the power of God.

After Kent Brantly and Nancy Writebol each got a course of ZMapp, there were four remaining courses in the world. A seventy-two-year-old Spanish priest named Miguel Pajares caught the virus in Liberia. He was airlifted to Madrid, where he was given one dose of ZMapp from Geneva Course No. 1, but he died soon afterward.

A British nurse named Will Pooley was working at the Kenema hospital as a volunteer. He broke with Ebola, and on August 24 he was flown to London in a plastic biohazard tent installed in the cargo hold of a Royal Air Force Boeing Globemaster transport aircraft. He was put in a biocontainment unit in a hospital in London and given the remaining two doses of Course No. 1. Pooley's Ebola disease turned around in twenty-four hours, and he recovered fully.

Larry Zeitlin sent Courses Nos. 4, 5, and 6 to ELWA Hospital in Monrovia, where they were administered to three African doctors who were sick with Ebola, named Abraham Borbor, Zukunis Ireland, and Aroh Cosmos Izchukwu. Dr. Borbor died but the others lived. At this point, Mapp Bio announced that all the available ZMapp in the world had been used up, all six courses of it.

The drug very clearly rescued Kent Brantly from imminent death. He was in the Ebola crash, and yet thirty minutes after ZMapp started hitting his bloodstream he sat up in bed. Within an hour, he was walking. How can this be possible? Only a very small amount of the drug had reached his bloodstream.

It seems that ZMapp swiftly killed every Ebola particle that was drifting in his bloodstream. It acted on Ebola the way a powerful insecticide freezes a nest of wasps. I asked Larry Zeitlin, the president of Mapp Bio, how it was possible for a very small amount of ZMapp to kill so many Ebola particles in just a few minutes.

Zeitlin wasn't surprised. He had seen antibodies do the same thing with sperm cells. After a bit of calculation, he said that thirty minutes after the drug began dripping into Brantly's arm, each individual Ebola particle in Brantly's bloodstream was surrounded with about thirty thousand individual antibodies—enough antibodies to nuke the particle and guarantee its death.

Brantly's blood was passing through his kidneys, and the kidneys are good at straining foreign particles out of the blood. Sixty minutes after Brantly was given ZMapp, he was in the bathroom peeing dead Ebola out of his body.

But that wasn't the whole story. Cells all through his body were factories squeezing out Ebola particles by the thousands. The antibodies also stuck to the Ebola hairs coming out of the cells, and killed the cells. Larry Zeitlin explained it this way: "With the caveat that we are in the area of total speculation—off the map, there be sea monsters here—there is a growing body of evidence that viral therapy with antibodies works by killing infected cells. This makes sense since you are stopping the factories from churning out virus rather than just killing whatever comes out of the factories." Bomb the factories that make the bombs, and you stop the bombs from being made. The patients needed three doses of ZMapp because the first dose wasn't able to kill every single infected cell. Some cells continued to pour out Ebola particles, but they got killed in the waves of ZMapp.

By early September, with the six official courses of ZMapp having been used up, Kentucky BioProcessing went into crash production of ZMapp, but the manufacturing process was very slow and yielded only small amounts of the drug. The secret course of ZMapp, though— Course Zero—remained unused. It sat in its freezer vault somewhere in the United States.

NATIONAL INSTITUTES OF HEALTH, BETHESDA, MARYLAND
Mid-August 2014

Officials at the National Institutes of Health combed through Lisa Hensley's electronic communications. "When they start looking at somebody's emails," Gary Kobinger said, "it's close to a done deal that the person's going to be fired." However, one fact quickly became clear. The NIH had *not* provided the drug to the American patients. Course No. 2 of ZMapp had been the property of the government of Canada. Gary Kobinger, a Canadian government scientist, had donated it to Samaritan's Purse for compassionate use in an American citizen.

Another fact came into play. Lisa Hensley had been on a deployment with the Department of Defense. She had been operating inside the military chain of command. The top CDC officer in Liberia, Dr. Kevin de Cock, had asked Hensley to review the best drug options for Samaritan's Purse: She had been subordinate to him in the chain of command. The U.S. ambassador to Liberia, Deborah Malac, had authorized the helicopter flight, and a Marine Corps officer, Lt. Col. Bryan Wilson, had asked Hensley to go on the flight as a representative of the U.S. government. The U.S. Ambassador has control over military people who serve in the Embassy, and Hensley's flight to get the ZMapp had been a diplomatic and military mission organized by the Embassy. One last thing was important. Two lives had been saved, and Hensley had made some small contribution to the effort. She hadn't gone rogue. Or had she? In a time of crisis, in the fog of a virus war, nobody is really in control. In any event, the investigation cleared Hensley of any wrongdoing. "Lisa did the right thing," Pardis Sabeti commented. The top management of the National Institutes of Health ultimately came to the same conclusion, and Hensley kept her job.

CAMBRIDGE, MASSACHUSETTS
September 22

On a warm day in the fall, Pardis Sabeti was in the Ebola War Room at the Broad Institute, running a meeting with a group of colleagues. Glass buildings of biotech and pharma companies filled the view in the windows, with the Charles River and Beacon Hill in the background. Humarr Khan had been dead for over a month. Ebola was continuing its expansion in West Africa. Nevertheless, it had been stopped in Nigeria. "The virus hasn't gone into exponential growth in Nigeria," Sabeti said to the group, "so we have a little bit of a respite."

Sabeti was certain that Ebola was soon going to show up in the United States, carried there by air travelers. She said that hospitals and health authorities weren't ready for the virus. Consequently, Americans were going to die of Ebola.

A week later, at Texas Health Presbyterian Hospital in Dallas, a man named Thomas Eric Duncan showed up in the emergency room with a headache, nausea, and a fever of 100.1. He'd been living in Monrovia, Liberia, in a rented room, and had recently arrived in the United States. After hours of tests, Duncan was given a prescription for antibiotics and discharged. Two days later he showed up at the same emergency room again, this time in an ambulance. Doctors learned, then, that he'd been living in Monrovia, which made them suspect he might have Ebola. They reported his case to the CDC.

Nurses and doctors at Texas Health Presbyterian Hospital gave care to Duncan while he was under suspicion of having Ebola. They got splashed with his body fluids, but they had not been wearing standard Tyvek biohazard suits, HEPA masks, or shoe coverings, though they had put on cotton surgical masks, gowns, gloves, and eye protection. In other words, they treated Ebola too casually. Duncan vomited on the floor, and the staff may have tracked Ebola particles around the hospital's corridors on their shoes. On September 29, a nurse named Nina Pham gave care to Duncan, and the medical records show no evidence that she wore any kind of protective gear. The CDC tested Duncan's blood and confirmed that he had Ebola; he died a week later.

Shortly after Duncan died, Nina Pham broke with Ebola. She was transferred to the biocontainment ICU at the NIH hospital in Bethesda, Maryland, for treatment, and survived. One of her principal physicians, who spent long hours at her bedside, was Anthony Fauci, the head of the National Institute of Allergy and Infectious Diseases. Another nurse from Texas Health Presbyterian Hospital, Amber Vincent, had also contracted Ebola from Thomas Eric Duncan. She traveled on a Frontier Airlines flight from Dallas to Cleveland while she was running an Ebola fever, and then returned to Dallas on another Frontier flight. Several hundred people were on those flights. Frontier Airlines deep-cleaned the planes four times and put the flight crews into isolation for twenty-one days. Nurse Vincent ended up at Emory University Hospital, and she also survived.

On Friday, October 17, a doctor named Craig Spencer, who'd been working as a volunteer for Doctors Without Borders in an Ebola treatment unit in Guinea, landed at Kennedy International Airport in New York, having finished his tour of duty. He felt deeply exhausted but otherwise okay. The next Tuesday, he walked along the High Line— a park that follows an elevated train track in Manhattan—and drank coffee at a café. He visited a food shop and ate meatballs, he rode the subway, he went bowling. On Thursday Dr. Spencer woke up feeling strange. His respiration was fast—he was breathing rapidly—and he was warm. By evening, a blood test had confirmed he had Ebola. New York City health officials were uncertain about how to biocontain him—cocoon him so he couldn't transmit the virus to anyone else in New York City. "It was clear they didn't have a plan," Spencer later said to *New York* magazine.

Spencer ended up isolated in a high-biocontainment unit at Bellevue Hospital, where he spent nineteen days with the virus, and ultimately recovered, having received top-notch care by a Bellevue medical team. The public and media got very nervous about how he had ridden the subway and gone many places in the city while the virus was in his body but he didn't have a fever. Spencer felt that public health authorities and the media had overblown the danger he'd posed to the city, and had unnecessarily frightened people by talking about whether or not you can catch Ebola from a bowling ball. In any event, Spencer's virus died inside the Bellevue biocontainment unit and didn't reach anyone else in New York.

In mid-October, the World Health Organization reported that there had been 9,200 cases of Ebola and 4,500 deaths. It was very clear that Ebola was growing explosively. The virus had reached a critical point. Epidemiologists had completely lost track of it in West African cities. Nobody—not doctors, not citizens, knew who had Ebola and who didn't. Ebola was spreading at a rate comparable to seasonal flu when it arrives in a city. As they projected the growing number of cases into the future, some epidemiologists predicted that there would be millions of Ebola cases within a year. The Ebola treatment centers were overwhelmed.

In Sierra Leone, the virus was raging in the capital and had spread throughout the country. There had been more than ten thousand cases reported, and the numbers were still climbing.

And yet something was happening in the Makona Triangle, at ground zero of the virus's emergence from the ecosystem. There, the number of new Ebola cases was dropping dramatically. By the end of October, there were almost no new cases of Ebola appearing in the Makona Triangle, and not many new cases around Kenema. Gradually at first, and then suddenly, the Makona strain faded and vanished from the cradle of its birth. Something very strange indeed had happened in the Makona Triangle.

THE CHAIN OF CARE

"At some point, people just got it," Lina Moses said, months later. The Kissi villagers in the Makona Triangle were the first to understand the truth: Ebola wasn't a fiction or a plot by foreigners, it was a communicable disease. People in the Makona Triangle learned the signs and symptoms of the disease. They avoided contact with anybody who looked like they might have the disease. They stopped going to funerals. In addition, they began sending their loved ones to the Doctors' camps. And eventually the same thing happened all over West Africa. "What they came to understand is that they cannot take care of the people they love," Lina Moses continued. "They have to surrender their infant, their spouse, or their beloved grandmother to an isolation ward in order to save the rest of their household. I think I would have a really hard time doing that in the same situation. When the stakes are your life and the lives of your family members, you figure things out pretty fast."

Anja Wolz, the Kailahun camp's clinical manager, made journeys to Kissi towns and villages in the Triangle during August and September of 2014, at the time when the virus was exploding in the cities. What she saw in the Triangle was that the Kissi villages had started practicing reverse quarantine, closing themselves off from outsiders to prevent the virus from entering the village. This is exactly what villagers did at Yambuku in 1976. "The villagers were quarantining themselves at the local level. Anybody who came into a village was checked beforehand," Wolz said. "They were checking who was coming into the village, who was sick." Villagers inspected Wolz and her driver for symptoms and made them wash their hands in chlorine before they were allowed into a village. At one village, Wolz and her driver were told they couldn't go in at all, because the village had isolated itself from the world. "It was like something out of the past," Wolz said. "They weren't going to burials, they stopped kissing each other, they weren't touching each other. Behavior changes."

"This is how all outbreaks end," Armand Sprecher, the Doctors' official in Brussels, said. "It's always a change in behavior. Ebola outbreaks end when people decide they're going to end."

In the cities, people wouldn't touch or go near any place or person that might have particles. When Ebola appeared in a family, the neighbors would shun the family. If you lived in Wellesley, Massachusetts, and a Level 4 virus was going around town, and somebody on your street was sick at home, you might not want your kids to play with those kids. Africans continued shunning Ebola-affected families even after the virus was gone from the family. People who wandered in the streets looking sick, or who lay on the streets dying, were left to die alone. There was no help for strangers. All across West Africa, people stopped shaking hands, stopped hugging or touching one another, and they went full OCD about washing their hands in bleach water. For a while, people changed their burial practices, too. No sane person wants to kiss an Ebola-ridden corpse.

The Ebola war wasn't won with modern medicine. It was a medieval war, and it went down as a brutal engagement between ordinary people and a life form that was trying to use the human body as a means of survival through deep time. In order to win this war against an inhuman enemy, people had to make themselves inhuman. They had to suppress their deepest feelings and instincts, tear down the bonds of love and feeling, isolate themselves from or isolate those they loved the most. Human beings had to become like monsters in order to save their human selves.

In West Africa there was no tradition similar to the Ancient Rule of the Congo Basin. But in 2014 the rules of the engagement with Ebola were exactly the same as the rules that Dr. Jean-François Ruppol had proposed to the people of Zaire when he stood on the table in the marketplace in 1976. The virus was contagious in the liquid humors of the body. If you could recognize the symptoms, if you didn't touch the liquids, if you avoided contact with people who had the symptoms, and if you let go of the dead, you could save yourself from infection.

By early October 2014, Monrovia was getting savaged by Ebola. All the beds in all the Ebola treatment centers were full, and there were simply no beds left for anybody sick with Ebola; people were caring for victims at home. Doctors Without Borders, in desperation, decided to distribute 65,000 Ebola disinfection and protection kits across Monrovia. The kits were simple, cheap, and primitive, and included a plastic bucket, bleach, a surgical gown, a mask, and gloves. Doctors' staff went around handing out the kits and advising people how and when to use them. A kit could be deployed for handling a dead body, or a caregiver could use it to protect themself while they were caring for a sick person.

In a town in Liberia, a young woman named Fatu Kekula, who was a nursing student, ended up caring for four of her family members at home when there was no room for them in a hospital—her parents, her sister, and a cousin. She didn't have any protective gear, so she created a bio-hazmat suit out of plastic garbage bags. She tied garbage bags over her feet and legs, put on rubber boots over the bags, and then put more bags over her boots. She put on a raincoat, a surgical mask, and multiple rubber gloves, and she covered her head with pantyhose and a garbage bag. Dressed this way, Fatu Kekula set up IV lines for her family members, giving them saline solution to keep them from becoming dehydrated. Her parents and sister survived; her cousin died. And she herself remained uninfected. Local medical workers called Fatu Kekula's measures the Trash Bag Method. All you needed were garbage bags, a raincoat, and no small amount of love and courage. Medical workers taught the Trash Bag Method, or variants of it, to people who couldn't get to hospitals.

Slowly at first, then more surely, the number of new Ebola cases began to drop. As the number of new cases dropped, the total number of Ebola particles in the swarm dropped at the same time. The particles weren't able to jump to fresh hosts, and the swarm began to shrink rather than grow. Trapped in the host they had killed, unable to reach a new host, vast numbers of particles died along with the ruined host. By the end of 2014, Ebola was fading away. In the Makona Triangle it was virtually gone.

Mapp Bio and Kentucky BioProcessing made three batches of pharma-grade ZMapp, and in April 2015, the NIH began a trial of ZMapp in Sierra Leone. By this time there were so few Ebola patients that it was very difficult to get statistics on ZMapp. In any event, the drug was given to eleven patients, and all of them recovered. But then the drug failed to save a boy, who died soon after receiving one dose of the drug. There just wasn't enough statistical evidence to say for sure that ZMapp is effective against Ebola. The U.S. Food and Drug Administration ruled that ZMapp showed promise but could not be licensed for emergency human use without more testing in animals.

Can ZMapp actually hammer down Ebola in minutes? The small amount of evidence, so far, suggests that ZMapp can, in fact, cure Ebola disease in some or many people, and that the drug can act extremely rapidly in some people. Antibody drugs have begun to look like a huge advance in medicine. Researchers, including the scientists at Mapp Bio, have been developing more antibody drugs against Ebola and for other viruses. ZMapp may be an example of a class of silver-bullet drugs—angels' swords—that could cure a person of many kinds of infectious diseases. Someday there could be antibody drugs that can cure infections by viruses, by drug-resistant bacteria, even possibly cure diseases caused by advanced bioweapons. Whether or not ZMapp proves to be a sure-fire Ebola-killer, it had been a huge breakthrough in the war on infectious diseases.

While Kentucky BioProcessing was rushing to make more ZMapp, there was only one pharmaceutical-grade course of the drug anywhere in the world. This was Course Zero, sitting in a freezer somewhere in the United States. When Course Zero was the only ZMapp in the world, it seemed extremely precious—a national asset.

At the White House, officials in charge of the security of the President of the United States carefully studied the medical case file on Kent Brantly and Nancy Writebol. "I've seen the file," Gary Kobinger explained later. "When you read it you say, 'Wow, this really works.'" Sometime in the fall of 2014, an official at the White House contacted an official at the National Institutes of Health and asked him if there was any ZMapp available for the White House. The NIH then learned of the existence of Course Zero. Course Zero ended up sitting in a freezer at an undisclosed location in or near Washington, D.C. Course Zero, which might have been used on Lisa Hensley if she'd come down with Ebola, was now reserved for the exclusive use of the President of the United States. Just in case Ebola ever comes to Washington.

As the Ebola epidemic died down, the NIH started testing the VSV-ZEBOV vaccine for Ebola, and it showed solid effectiveness. At this writing, the vaccine is being tested in eastern Congo, where Ebola virus has broken out and is running wild—and no doubt evolving in human bodies.

As the tide of the Makona strain receded, it left its dead scattered across eight countries, including Spain and the United States. Thirty thousand people had been infected. More than eleven thousand people had died of the virus, and untold thousands more had died because they couldn't get medical care during the epidemic, since hospitals were devastated. Seven percent of all the doctors in Sierra Leone were dead. The medical infrastructure of Guinea, Liberia, and Sierra Leone had been wrecked. The three nations' economies had functionally collapsed. At Kenema Government Hospital, at least thirty-seven nurses were dead. Two Kenema hospital doctors were also dead—Humarr Khan and Sahr Rogers. In the end, the Ancient Rule prevailed, and the emerging virus temporarily went back to its hiding place in the virosphere.

STONY BROOK, LONG ISLAND, NEW YORK
About 1:00 p.m., June 1, 2016

In the spring of 2016—a year after the great Ebola epidemic subsided, a postdoctoral researcher named William "Ted" Diehl began doing experiments with Ebola proteins that had been collected from the different mutant Ebolas that evolved in the swarm as the epidemic went along. Diehl was looking closely at the different kinds of "fish" in the school, as it were, and was working in the lab of a prominent AIDS researcher, Dr. Jeremy Luban, at the University of Massachusetts Medical School.

Ted Diehl discovered that one of the Ebolas, the one now known as the A82V Makona Variant, or the Makona strain, was four times better able to infect human cells in a test tube than the wild, natural Ebola that got into the little boy in Meliandou. The Makona strain was really hot in human cells. Was there something different about this Ebola? What, exactly, made the Makona so hot?

Diehl's boss, Jeremy Luban, called Pardis Sabeti. He didn't know anything about the Makona strain other than it seemed very hot. Sabeti got excited. She told him that the Makona strain had taken over in West Africa. It had knocked out all the other Ebolas; it was the Ebola that had swept through the cities of West Africa, the one that had killed Humarr Khan, the one that had gotten to Dallas and New York. Yes, she told Luban, this Ebola was the fish with the sharpest teeth, the real killer in the swarm. And it was different from the other Ebolas by only *one* letter in its 18,959 letters of code. The change in one letter had caused a slight change in one of the proteins of Ebola.

Nobody knew what made the Makona strain hot. But on the first of June, 2016, Ted Diehl was in Stony Brook, Long Island, sitting at the dining room table in his wife's apartment, and drinking green tea. (The couple worked in different places.) He was looking at an image on his laptop of the exact structure of a certain protein that exists in the soft knobs that stud the outside of an Ebola particle. The knobs help an Ebola particle get inside a human cell.

Proteins are made of long strings of amino acids, which are like necklaces, and the necklaces are folded in special ways. In the case of the hot Ebola, the Makona strain, *one* amino acid was different in the protein of the soft knob. In the wild Ebola, the Meliandou Ebola, the amino acid was alanine. It got changed to a different amino acid, valine, in the Makona strain. The change seemed insignificant: Why should it make Ebola four times more infective?

Ted Diehl began rotating the image of the protein on his computer screen, studying its curious shape. All of a sudden, in a flash of insight, he saw that the shape of the protein could fit better on something that sticks out of a human cell membrane. Like a key fitting in a lock. He saw that the mutant Ebola protein could stick to the skin of a human cell better, and could open the cell up, so that the Ebola particle could get inside the cell more easily. The mutant Ebola sticks to a receptor knob on the outside of a human cell that pulls cholesterol into the cell, called the Niemann-Pick receptor. Ebola uses the Niemann-Pick receptor to invade human cells. (Niemann-Pick disease is an inherited, fatal disease in which a person's cells can't absorb cholesterol properly. For this reason, a person with Niemann-Pick disease is presumably immune to Ebola.)

Sitting there at the kitchen table that day, Ted Diehl became the first person to see exactly what gave the Makona strain an edge at invading humans. "It felt like holding a lottery ticket and seeing all the numbers come up," Diehl says. "Out of dumb luck we scored a huge win." Diehl had gotten a look into the microscopic halls of nature and had seen a tiny thing that looked different to him. Maybe he had seen how close the world came to something much worse than ten thousand deaths and three nations wrecked.

At the same time that Ted Diehl was figuring out what made the Makona strain so dangerous, a researcher named Jonathan K. Ball, at the University of Nottingham, in England, and colleagues of his, discovered something equally disturbing about the Makona strain. It *didn't* infect bat cells as well, while it infected human cells *much better*. In other words, the Makona strain is a humanized Ebola. The Makona strain understands people better than does any other kind of Ebola.

Pardis Sabeti had this to say: "The mutation increased the virus's ability to infect human cells, while it made the virus less able to infect other animals. As the virus was transmitting from human to human it was improving its ability to do so. We know viruses mutate. Most mutations of a virus don't do anything. But if you give the virus enough chances, a match may light, and a spark may go."

In other words, if the Makona strain hadn't been stopped quickly, it would have continued improving its ability to spread in humans. It would have become yet more humanized. The world got lucky this time. If the Makona strain had raced into a poor supercity, it would have gotten into many more thousands of people, and gotten many more chances to evolve and change. For a long time after the Ebola epidemic subsided, nobody really understood just how close the world had come to a much bigger disaster. What might have happened if the Makona strain had blown up in the supercity of Lagos, population twenty million, after Patrick Sawyer brought the virus to the city? If Lagos had gone hot, could the virus have moved to other cities around the world, and could more cities have gone hot? If the Makona strain had kept on evolving, getting to know the human body and the human immune system better and better, the Ancient Rule eventually could have arrived on the streets of Los Angeles, Tokyo, the industrial Ruhr of Germany, the shanties of São Paulo. We are one species, all connected, but we are just one thing to a virus: a host.

What would it be like if a Level 4 virus event occurred and the Ancient Rule arrived in the supercity of New York? It wouldn't take much to produce the Ancient Rule in New York City. A dry virus with high mortality that infects people through the lungs. No vaccine, no medical treatment for the virus. If you take the subway, if you ride in an elevator, you can be infected, too. If the Ancient Rule came to New York City, we can imagine people lying facedown on the street or in Central Park, crowds staring and hanging back. People begging for help, no one willing to help. Police officers wearing full PPE gear. People needing ambulances. No ambulances. Hospitals gone medieval. Medical staff absent, dying, overwhelmed. All hospital beds full. People being turned away on the street from Bellevue Hospital. Medical examiner facilities gone hot as hell and crammed with corpses. Nobody in their right mind would enter a New York City hospital during a time of the Ancient Rule. Transportation frozen. Food supplies dwindling or absent. Schools closed. People avoiding supermarkets for fear of contagion. Prophets and visioners predicting the future and offering cures. People leaving the city, bringing the virus with them. Airports infective, flights canceled. Parents giving care to their sick children in apartments, at home. If someone in a family got sick, there would have to be one designated caregiver, a person willing to sacrifice their life in an attempt to give care to a loved one. Wealthy people spending money like water trying to save themselves; the poor and disadvantaged, as always, bearing the worst of it. If there is a vaccine or drug that can help, there will be corruption. Companies and individuals hoarding vaccine, selling it at sky-high prices.

Pardis Sabeti spends her life studying how viruses evolve and change. For years, she has been telling her Harvard and Broad Institute colleagues that they should keep a month's supply of preserved food and basic medical supplies in their apartments or homes. A simple precaution, just in case of a Level 4 event. Just in case you might have to practice reverse quarantine for about a month, the same way African villagers do. Cut yourself off from the outside world for a time. "I want to set up the possibility that you might have to stay indoors for a while," she says to her staff.

"If we did some basic preparation for a major outbreak," Sabeti said recently, "we could actually make it not such a huge, dramatic, crazy thing." Sabeti refers to a Level 4 pandemic as a bananas event. "Why should we be waiting for something that's truly bananas to break out *before* we start planning for it?" she asked. "There's not a lot of value in preparing for a war, because what happens in a war is unpredictable. But there is a lot of value in preparing for an outbreak, because what happens in an outbreak *is* predictable. Let's be prepared, not scared."

• • •

There is one more story to tell about Kenema, terrible and biblical in its simplicity. In the summer of 2017, Pardis Sabeti and Robert Garry delivered a genome sequencing laboratory to the Lassa research program at Kenema Government Hospital, and the local staff was trained in genome sequencing. During 2018 and 2019, Sierra Leonian technicians sequenced the exact Ebolas that had been in the blood of Kenema Ebola patients. In other words, they revealed the genetic code of the virus swarm that murdered their friends and medical colleagues. The code didn't lie. It revealed an invisible history, shocking and deeply moving.

Most observers assumed that the Kenema nurses caught the virus from patients or from people in the local community. The letters of the code told a different truth. The Kenema staff *caught the virus from one another* as they tried to save one another's lives. A catastrophic chain of infections among the staff began from a small event. On or about June 30, 2014, ambulance driver Sahr Nyokor decided to break a rule. He didn't want to embarrass himself or frighten people by wearing a moon suit when he visited some friends in their house. He wore no protection when he entered the house. Somebody inside the house was sick with the Makona strain. That person was closely connected to the funeral of Menindor. The hot Makona jumped from that person into Mr. Nyokor. He threw up blood and went to the hospital. At around six o'clock in the morning on June 18, Mr. Nyokor fell in the toilet and cut his scalp. Nurse Lucy May washed the blood from his cut—a gentle act of care—and he died soon afterward. A few particles of Mr. Nyokor's hot Makona got into Lucy, likely when she cleaned his scalp, and began explosive amplification in her fetus and herself.

The Makona strain jumped out of Lucy May into her caregivers on the night of July 3 while she died giving birth to a stillborn child. The code didn't lie: Auntie Mbalu Fonnie caught the virus from Lucy May as she desperately tried to save Lucy's life while performing an abortion of the dead baby. The three nurses, Princess Gborie, Sia Mabay, and Fatima Kamara, all caught the virus from Lucy May, most likely on that same night while they, too, tried to save Lucy's life. Nurse Alex Moigboi also caught Lucy May's virus while he gave her tender care during the night shifts. It seems pretty clear that Auntie knew she could die if she attempted to save Lucy May. The nurses knew it, too. Auntie and her nurses made the ultimate sacrifice as they attempted to rescue Lucy May; they were like the firemen who ran into the World Trade Center moments before the tower collapsed. They did their duty because it was what they had to do. All of them caught Ebola from Lucy May as they attempted and failed to rescue her from death. Sia Mabay and Fatima Kamara survived their ordeals; Princess Gborie perished along with Auntie.

Lucy May's Makona strain expanded out of Auntie. Her brother, Mohamed Yillah, caught Lucy May's virus from his sister as he tried to save her life.

And what about Humarr Khan? The letters of the code showed that he, too, was consumed by the fire that started when the ambulance driver went into somebody's house. The code shows that Khan died of Lucy May's virus. Wherever and however Dr. Khan caught Lucy May's virus, he was infected while he was giving care to his own staff. The code shows that Khan did not fail his people, he died with them.

The virus, a true monster, followed the bonds of fealty and love that joined the hospital's caregivers to one another and ultimately to every other person on earth. The African medical professionals gave their lives trying to rescue one another, and, at the same time, they served as a thin, dissolving line of sacrifice in which they stood between the virus and you and me.

EPILOGUE
A Level 4 Event

The past is unpredictable. When I first started researching this book, right at the time Humarr Khan died, I had no idea what I would find or where the story would go. I have done my level best to make this narrative accurate and faithful to the strange twists and turns of reality, a bricolage of events, as time goes by. In my view, no work of invented fiction can quite approximate a sense of random coincidence that feels simultaneously like fate. Now, as the past moves into the future, I propose to look ahead. There is a caveat: I may not be any better at prophecy than Wahab the Visioner, although he did make at least one good call. What I propose to look at is a global outbreak that might be termed a Level 4 event worldwide outbreak of a Biosafety Level 4 emerging virus that travels in the air from person to person, and is vaccineless and untreatable with modern medicine.

The Ebola epidemic seems to be part of a pattern rather than something unusual or extreme. When looked at closely, it was really just a series of small accidents and unnoticed events, which, moment by moment, grew into a crescendo of horror. This was the shockwave produced by an emerging virus as it came out of the ecosystem. The virus magnified itself in people, swept away lives, met opposition from the human species, and finally died out. What will the next shockwave be like?

As far as anyone knows, Ebola doesn't travel through the air from person to person. Ebola is a wet virus, and it spreads through contact with liquids or in invisible liquid droplets that can drift a few feet through the air. The question often asked is whether Ebola could evolve to spread through the air in dried particles, entering the body along a pathway into the lungs. Eric Lander, the head of the Broad Institute, thinks that this is the wrong question to ask. Lander is tall, with a square face and a mustache, and he speaks rapidly and with conviction. "That's like asking the question, Can zebras become airborne?" he said. In order to become fully airborne, Ebola virus particles would need to be able to survive in a dehydrated state on tiny dust motes that remain suspended in the air and then be able to penetrate cells in the lining of the lungs. Lander thinks that Ebola is very unlikely to develop these abilities. "That would be like saying that a virus that has evolved to have a certain lifestyle, spreading through direct contact, can evolve all of a sudden to have a totally different lifestyle, spreading in dried form through the air. A better question would be 'Can zebras learn to run faster?'" There are many ways by which Ebola could become more contagious even without becoming airborne, Lander said. For example, it could become less virulent in humans, causing a milder disease and killing maybe 20 percent of its victims instead of 50 percent. This could leave more of them sick rather than dead, and perhaps sick for longer. That might be good for Ebola, since the host would live longer and could start even more chains of infection. But Ebola will probably always be a wet virus.

As this is being written, Ebola has become another shockwave, this time in eastern Democratic Republic of the Congo, where hundreds have died of Ebola and the virus is out of control. Nobody knows if or how Ebola might mutate as it chains through human bodies in the current outbreak. But let us suppose that Ebola never evolves into a more serious problem than it is right now. Instead, let us consider a dry virus. Dry viruses have the ability to survive when their particles have no moisture in them. The particles, stuck to dust motes or in microscopic dry flakes of saliva, can drift in the air for long distances.

• • •

A family of viruses called the morbilliviruses is regarded by some experts as a leading candidate for the emergence of a previously unknown Level 4 monster that travels in the air. If there was no vaccine or drug for it, and if it was highly infectious, and if it floated out of peoples mouths, the virus could go around the world in a few weeks, traveling inside people who are flying on airplanes and walking through airport terminals, breathing. Consider a certain morbillivirus, Nipah, a Level 4 emerger that gets into the lungs and central nervous system, and causes personality changes and liquefaction of the brain. Nipah occasionally breaks out in southeast Asia. It jumps from bats, to animals, to humans—it is a promiscuous virus. Nipah isn't very contagious right now, but viruses evolve in response to people. And there are other Nipah-like viruses circulating in living creatures in the world's ecosystems. If a brain-destroying virus was going around like the flu, every person's risk factor for catching the virus would be breathing. If you live in a city teeming with humans, you are nothing more than a host.

Mapp Biopharmaceutical has now created an antibody superdrug for Ebola called the Pan-Ebolavirus Cocktail. This new drug is effective against all species of Ebola virus. As this is being written, the U.S. government is preparing to issue funding to manufacture a huge supply of Pan-Ebolavirus Cocktail to be stored in the Strategic National Stockpile, a secret facility or facilities for mass storage of drugs and vaccines to protect the population against biological weapons and emerging viruses. There is also a new antibody drug for Nipah virus. The Pan-Ebolavirus Cocktail and the Nipah Cocktail are models for drugs of the future—antibody drugs that might be developed quickly in a global emergency and that can be surge-manufactured fast, in large quantities. This is the future. And yet right now we are not prepared.

A recent study done at the Johns Hopkins University School of Public Health revealed that in all the hospitals in the United States there are only a total of 142 biocontainment red zone beds for patients with a hemorrhagic fever virus such as Ebola. And there are not more than 400 red zone beds for patients infected with an airborne hot virus. So there are a total of 542 hospital beds available in the United States in case of a Level 4 event. And there is a big question whether there are even enough trained nurses and doctors to care for the patients in those 542 red zone beds.

A question has to be asked: If a Level 4 emerging virus spread to a million people in North America, or in any continent, would hospitals be able to handle the patients and give them care? Would epidemiologists be able to trace and break the chains of transmission if a million people were infected?

As I think about the supercities of the earth, an image comes to my mind of a field of storage tanks full of fuel. All the tanks are connected by pipes carrying fuel, and the valves in the pipes cannot be closed off completely. If one tank goes up in flames, the entire field of tanks can blow. What the future may hold for the human species in its relationship with the virosphere is in the realm of human choice and the play of chance. Wahab the Visioner believed that we can change our fate if we can see it coming, and that human actions can sometimes, not always, change a pattern of events as they jiggle and bump their way into the future.

By now, the warriors who stand watch at the gates of the virosphere understand that they face a long struggle against formidable enemies. Many of their weapons will fail, but some will begin to work. The human species carries certain advantages in this fight and has things going for it that viruses do not. These include self-awareness, the ability to work in teams, and the willingness to sacrifice, traits that have served us well during our expansion into our environment.

If viruses can change, we can change, too.

GLOSSARY

amplification Strong multiplication of a **virus**, leading to a large increase in the number of virus particles. See also **replication.**

Biosafety Level 4 Also BSL-4 or Level 4. Highest level of **biocontainment**; requires the wearing of a bioprotective space suit.

biocontainment The methods and technology for containing a hazardous biological agent and preventing it from infecting people.

biosphere The totality of the global ecological system of all living organisms. See also **virosphere.**

cadaveric blood Blood from a corpse.

chain of transmission The movement of an infectious agent going from person to person.

cross-species jump of a virus The process whereby a virus changes the type of host it infects, jumping from one kind of host to a different kind of host.

electron microscope A powerful microscope that uses a beam of electrons to make an image of something very small.

emerging virus "Viruses that have recently increased their incidence [in humans] and appear likely to continue increasing." Term and definition coined by virologist Stephen S. Morse. Many emerging viruses exist naturally in animal hosts and make **cross-species jumps** into humans.

epidemiology The science and art of tracing the origin and spread of diseases in populations, with the goal of controlling or stopping the diseases.

filovirus Family of viruses that are genetically related and all have a similar threadlike or stringlike shape.

genome "The complete set of genes or genetic material present in a cell or organism." (Oxford Dictionaries.)

host An organism that a parasite lives in or on.

hot agent A **Biosafety Level 4** virus.

hot Virulent (causing severe disease) and highly infectious.

mutation A change in the "spelling" of the genetic code of an organism, which can sometimes result in a change in the biology and character of the organism.

parasite An organism that lives inside or on a **host** organism and harms the host or does it no good.

pathogen A disease-causing microbe or **virus**.

PCR machine A machine that uses the polymerase chain reaction to detect genetic code in a sample such as a blood sample.

PPE Personal protection equipment. Nonpressurized bioprotective PPE typically consists of an impermeable suit that covers the whole body from head to feet, eye protection, a high-efficiency breathing mask, protective gloves, and rubber boots.

red zone Extreme biocontainment ward for designed for isoation of patients infected with a highly dangerous **virus**.

replication Self-copying. See also **amplification**.

virosphere The totality of the system of **viruses** in the world of living nature. See also **biosphere**.

virus Very small replicating life form, and parasite, consisting of a capsule made of proteins which contains DNA or RNA, which is the virus's genetic code. A virus is only able to replicate inside cells of a host.

ACKNOWLEDGMENTS

Many people provided important assistance for the making of this book. I am eternally grateful to them, because without their kind help the book wouldn't exist. If I've left out someone's name in the following list, the omission is inadvert, and I do apologize for it. Any errors of fact in this book are mine alone.

Kenema Government Hospital

First and foremost, I wish to thank the staff of Kenema Government Hospital for their generous help in sharing with me their recollections, thoughts, and perceptions, as well as for their unfailing kindness to me, a visitor. I can only hope to use my best words to do some sort of justice to their courageous and dedicated service in medicine and public health. I wish to specially thank Francis Baimba, James Bangura, Gabriel Bundu-Kainessie, Mohamed Fomgbeh, Michael Aiah Gbakie, Augustine Goba, Dr. Abdul Azziz Jalloh, Simbirie Jalloh, Fatima Kamara, Lansana Kanneh, Veronica Jattu Koroma CN, James Koninga, Mambu Mohmoh, Doris Moriba, Joseph Henry Moseray, Isaac Tucker Musa, Ibrahim Saffa Ngobah, John Sesay, Mohamed Sow, Dr. Mohamed A. Vandi, and Mohamed Sankoh Yillah.

Tulane University

Robert F. Garry, Jeneba Abu Kanneh, Dr. John S. Schieffelin, Sheku Show.

Harvard University and the Broad Institute

Michael J. Butts, Andrew Hollinger, Daniel J. Park, Pardis Sabeti, Sarah Winnicki, Nathan Yozwiak.

National Institute of Allergy and Infectious Diseases— Integrated Research Facility

Lisa E. Hensley, Dr. Anthony S. Fauci, Anna Honko, Peter B. Jahrling, Curtis Klages, DVM, Dr. Jens H. Kuhn, Mark Martinez, DVM, Gene Garrardy Olinger, Jr. Also: James Hensley, Dr. Mike Hensley, Karen Hensley.

Doctors Without Borders/Médecins Sans Frontières

Dr. Bertrand Draguez, Dr. Armand Sprecher, Anja Wolz RN.

Other Individuals

Kristian G. Andersen (Scripps Research Institute), Dr. Daniel G. Bausch (UK Public Health Rapid Support Team), Dr. Joel G. Breman (Centers for Disease Control and Prevention), Alexander Bukreyev (Galveston National Laboratory), William "Ted" Diehl (University of Massachusetts Medical School), Dr. Joseph Fair (Fondation Mérieux USA), Dr. Tom Fletcher (Liverpool School of Tropical Medicine), Dr. Robert Fowler (Sunnybrook Research Institute), Thomas W. Geisbert (Galveston National Laboratory), Stephen Gire (NexGen Jane Inc.), Frédérique A. Jacquerioz (University Hospitals of Geneva), Macmond M. Kallon (Government of Sierra Leone), Gary P. Kobinger (CHU de Quebec-Université Laval), Dr. Thomas G. Ksiazek (Galveston National Laboratory), James LeDuc (Galveston National Laboratory), Fabian Leendertz (Robert-Koch Institut), Jeremy Luban (University of Massachusetts Medical School), Frederick A. Murphy (formerly of the CDC and Galveston National Laboratory), Dr. Jean-Jacques Muyembé-Tamfun (D.R. Congo National Institute for Biomedical Research), Michael T. Osterholm (University of Minnesota), Dr. Peter Piot (London School of Hygiene and Tropical Medicine), Dr. Lance Plyler (Samaritan's Purse), Dr. Jean-François Ruppol,

Josiane Wissocq, Erica Ollman Saphire (La Jolla Institute for Immunology), Randal J. Schoepp (USAMRIID).

The Khan Family

The late Ibrahim Seray Khan, Ms. Aminata Khan, Alhajie Alpha Khan, C-Ray Khan, Sahid Khan.

Professional

The editor of this book is Mark Warren of Random House; he did an amazing job editing a complex manuscript that mutated faster than Ebola. Also at Random House: Carlos Beltrán, Melanie DeNardo, Susan Kamil, Tom Perry, Chayenne Skeete, and Andy Ward. Many thanks to Lynn Nesbit and Cullen Stanley. Bruce Vinokour of Creative Artists' Agency gave me an important suggestion that strengthened this book. And many thanks to Dr. Gary Karpf.

I wish to thank certain individuals for very special help: Eric S. Lander (Broad Institute) for an act of kindness; Lina M. Moses (Tulane University) for help understanding Sierra Leonian culture and for deepening my understanding of key characters in this narrative; Nadia Wauquier (MRI Global) for her precise and eloquent accounting of many details of the Kenema outbreak; Dr. Jean-Louis Lamboray (La Constellation) for his public health perspective and for his assistance during fascinating interviews with Dr. Jean-François Ruppol; and Sarah Claus Butler for her help as a translator during interviews with Dr. Ruppol.

Love and deepest gratitude to my wife, Michelle. She provided centrally important editorial guidance for this book. She also gave me inestimable love and support during the long process of research and writing. Our three children, Oliver, Laura, and Marguerite, themselves writers, have continually inspired me with their creativity and wisdom.

PHOTO © ROBERT LEWIS

RICHARD PRESTON is the *New York Times* bestselling author of ten books, including *The Hot Zone*, *The Wild Trees*, and *The Demon in the Freezer*. Preston has taught nonfiction writing at Princeton University and the University of Iowa, and he is the recipient of many prizes and honors, including the Champion of Prevention Award of the U.S. Centers for Disease Control. His books have been published in more than thirty languages.